1st Grade
Math Workbook

Addition and Subtraction

Want Free Extra Goodies for Your Student?
Email us at: info@homerunpress.com.
Title the email "1st Grade Math Workbook"
and we'll send some extra worksheets your way!

We create our workbooks with love and great care.
For any issues with your workbook, such as printing errors, typos, faulty binding, or something else, please do not hesitate to contact us at: info@homerunpress.com.
We will make sure you get a replacement copy immediately.

THANK YOU!

First published in the USA 2020.   ISBN 9781952368042

Table of Contents

Hi. I'm Sunny. For me, everything is an adventure. I am ready to try anything, take chances, see what happens - and help you try, too! I like to think I'm confident, caring and have an open mind. I will cheer for your success and encourage everyone! I'm ready to be a really good friend!

I've got a problem. Well, I've always got a problem. And I don't like it. It makes me cranky, and grumpy, impatient and the truth is, I got a bad attitude. There. I said it. I admit it. And the reason I feel this way? Math! I don't get it and it bums me out. Grrrr!

Not trying to brag, but I am the smartest Brainer that ever lived - and I'm a brilliant shade of blue. That's why they call me Smarty. I love to solve problems and I'm always happy to explain how things work - to help any Brainer out there! To me, work is fun, and math is a blast!

I scare easily. Like, even just a little ...Boo! Oh wow, I've scared myself! Anyway, they call me Pickles because I turn a little green when I get panicky. Especially with new stuff. Eek! And big complicated problems. Really any problem. Eek! There, I did it again.

Hi! Name's Pepper. I have what you call a positive outlook. I just think being alive is exciting! And you know something? By being friendly, kind and maybe even wise, you can have a pretty awesome day every day on this amazing planet.

A famous movie star once said, "I want to be alone." Well, I do too! I'm best when I'm dreaming, thinking, and in my own world. And so, I resist! Yes, I resist anything new, and only do things my way or quit. The rest of the Brainers have math, but I'd rather have a headache and complain. Or pout.

1. Read.

When I add ③ candies and ② candies, there are ⑤ candies altogether. It does not matter which way I add candies together.

3 + 2 = 5 candies

means equals   means add or plus

2 + 3 = 5 candies

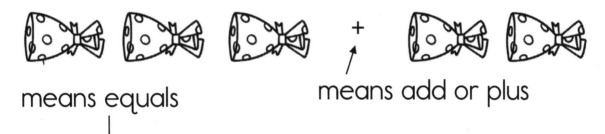

I add ④ cars and ③ trucks.

I have 4 cars and 3 trucks together. I can find the total simply by counting them all. There are 7 in all.

4 + 3 = 7

1. Read.

I use a number line to find out the answer when I add 4 and 2. First, I <u>draw</u> a line and <u>mark</u> it with numbers. I find 4 on the number line.

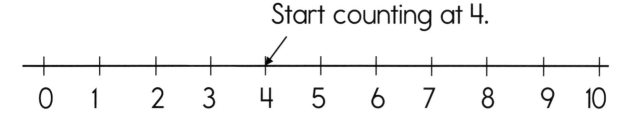

Start counting at 4.

I need to add 2, so I jump 2 places to the right.

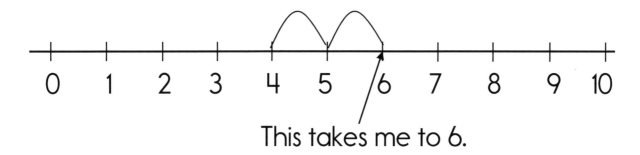

This takes me to 6.

So $4 + 2 = 6$

I add 30 and 50. First, I find 30. Then, I jump 5 places to the right.

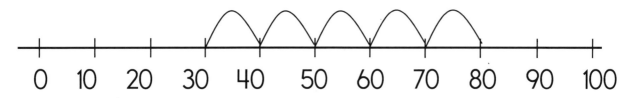

So $30 + 50 = 80$

www.homerunpress.com

1. <u>Add</u>. Use a number line to show the jumps.

2 + 7 = ___

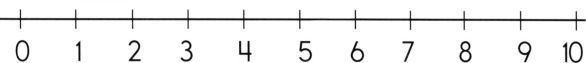

6 + 4 = ___

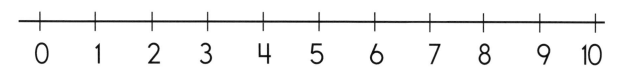

9 + 1 = ___

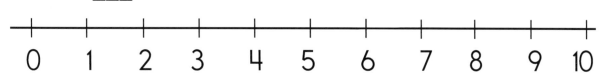

50 + 50 = ___

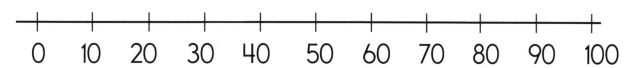

40 + 40 = ___

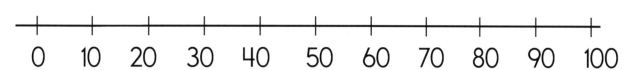

30 + 70 = ___

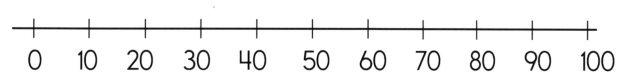

1. <u>Compare</u>, using ">," "<," or "=." The first one is done for you.

3 + 7 > 10 – 2		8 – 5	4 – 1
4 + 2 9 – 2		7 - 2	10 – 6
2 + 5 10 – 2		8 – 5	10 – 7
6 + 3 4 + 3		6 – 4	5 – 2

"We need to add 3, 4 and 2". <u>What</u> is the total?

2 + 3 + 4 = _

3 + 4 + 2 = _

4 + 3 + 2 = _

The order of addends does not affect the total – the sum will be the same.

Aha, I remember something about that. I've read that rule called the COMMUTATIVE property.

That's true. 2 candies + 3 candies or 3 candies + 2 candies = always 5 candies!

2. <u>Solve</u> the puzzle.

 + + +  = 10

 =

 + + = 7

 =

1. <u>Add</u> and <u>change</u> the addends' order. The first one is done for you.

2 + 5 + 3 = 10   5 + 3 + 2 = 10   3 + 2 + 5 = 10

1 + 4 + 2 = _   _ + _ + _ = _   _ + _ + _ = _

5 + 0 + 5 = _   _ + _ + _ = _   _ + _ + _ = _

4 + 1 + 3 = _   _ + _ + _ = _   _ + _ + _ = _

6 + 1 + 3 = _   _ + _ + _ = _   _ + _ + _ = _

2. <u>Complete</u> an addition number sentence with tens and ones.

16 = 10 + 6   18 = __ + __   15 = __ + __

10 = __ + __   13 = __ + __   17 = __ + __

3. <u>Write</u> the missing numbers to make the comparison true.

1 + 9   =   _ – 4   9 – 4   =   _ – 2

3 + 4   =   10 – _   10 - _   =   1 + 3

4 + 4   =   _ + 2   _ – 5   =   8 – 7

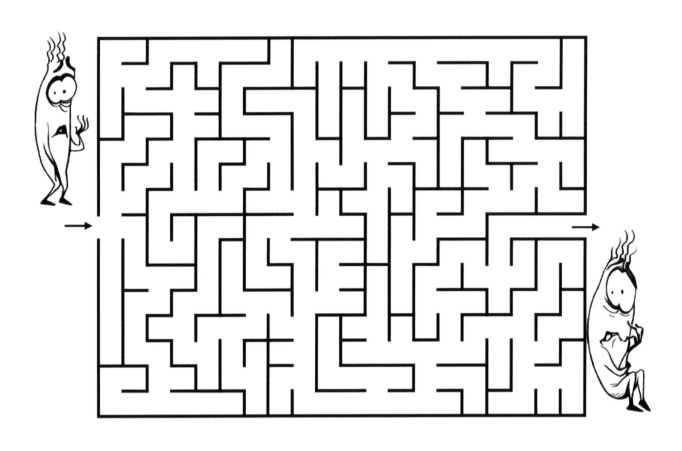

1.

I'm faster than Pickles, but Sunny is faster than me. <u>Who</u> is the fastest?

Answer:

_____

1. The blocks in each tower tells you how many hundreds, tens, and ones in each number. <u>Write and put</u> the numbers in order from the least to the greatest.

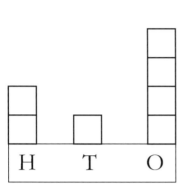 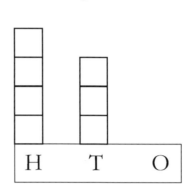 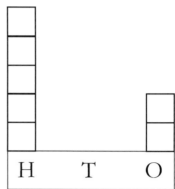

_____   _____   _____

2. <u>Write and put</u> the numbers in order from the largest to the smallest.

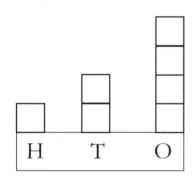 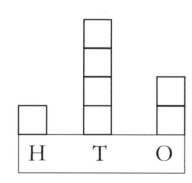 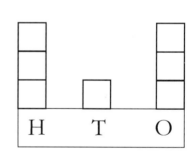

_____   _____   _____

3. My sister found 9 apples. She ate 6 apples. <u>How many apples</u> are left?

Answer: _____

1.   Make and write the smallest and the biggest two-digit numbers you can with any two of these digits: 1, 7, 4, 6.

Answer: _____

2.   Make and write the smallest and the biggest two-digit numbers you can with any two of these digits: 9, 5, 9, 8.

Answer: _____

3. Make and write the smallest and the biggest two-digit numbers you can with any two of these digits: 2, 9, 3, 0.

Answer: _____

4. Write the missing number.

___ + 40 + 8 = 548   100 + ___ + 6 = 126

200 + 50 + ___ = 251   ___ + 80 + 9 = 389

___ + 10 + 0 = 610   400 + ___ + 4 = 484

700 + 30 + ___ = 737   ___ + 60 + 2 = 862

___ + 10 + 1 = 911   500 + ___ + 1 = 541

300 + 60 + ___ = 369   ___ + 70 + 2 = 572

1. Read.

I like to split the adding numbers into numbers that are easier to work with. I can show my favorite strategy. T = tens, O = ones.

Step 1. Let's add 12 and 15.

T O	T O
12 + 15 = ___	

Step 2. Add the tens together.

T O	T O	T O
10 + 10 = 2 0		

Step 3. Add the ones together.

T O	T O	T O
2 + 5 = 7		

Step 4. Add the tens and ones to find the total.

T	O	T O
20 + 7 = 2 7		

1. Add.

16 + 12 = 10 + 10 + 6 + 2 = ___ + ___ = ___

21 + 17 = ___ + ___ + __ + __ = ___ + ___ = ___

13 + 15 = ___ + ___ + __ + __ = ___ + ___ = ___

I can add numbers using column addition. Hint: Write ones under ones. Write tens under tens.

Step 1: Write the digits that have the same place value lined up one above the other.

tens ones

```
    1   2
+   1   5
  ─────────
    -   -
```

Step 2: Start by adding the ones together. Add 2 ones and 5 ones: 2 + 5 = 7 Write 7 in the ones column.

tens ones

```
    1   2
+   1   5
  ─────────
        7
    -
```

Step 3: Add 1 ten and 1 ten. But I actually add 10 and 10. So the answer is: 10 + 10 = 20 I write 2 in the tens column.

tens ones

```
    1   2
+   1   5
  ─────────
    2   7
```

1. Add.

```
    3         5         3         7         4         8
  + 3       + 2       + 6       + 2       + 4       + 1
  ─────     ─────     ─────     ─────     ─────     ─────
    6
```

```
    2         5         3         4         1         6
  + 8       + 5       + 7       + 6       + 9       + 4
  ─────     ─────     ─────     ─────     ─────     ─────
```

2. Complete each pair of number bonds.

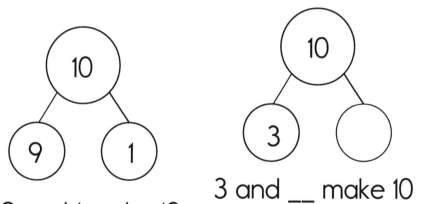

9 and 1 make 10

3 and __ make 10

7 and __ make 10

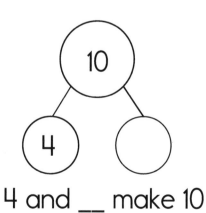

2 and __ make 10

4 and __ make 10

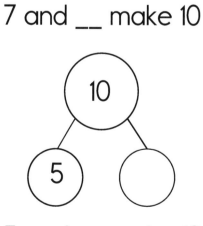

5 and __ make 10

1. <u>Use</u> cupcakes to make 10. <u>Color</u> the cupcakes brown and yellow.

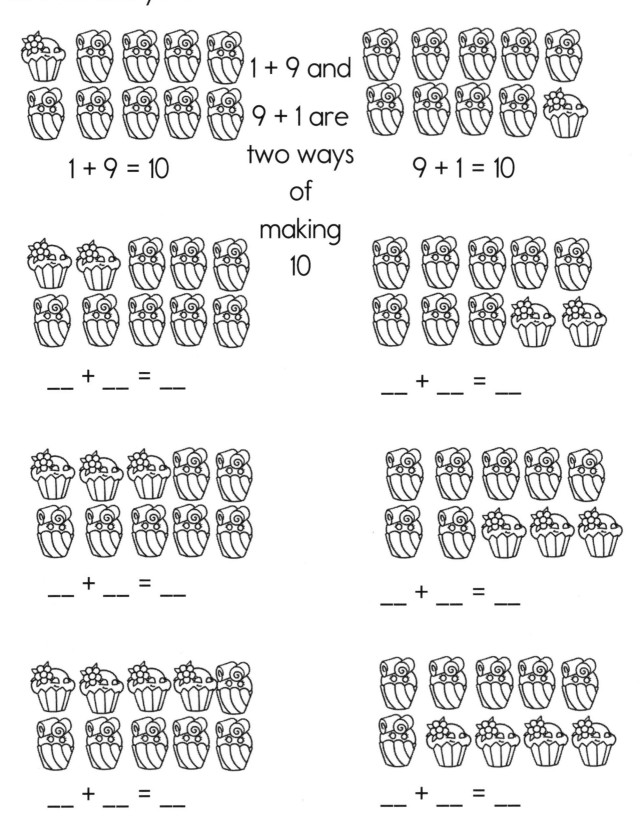

1 + 9 = 10

1 + 9 and 9 + 1 are two ways of making 10

9 + 1 = 10

__ + __ = __

__ + __ = __

__ + __ = __

__ + __ = __

__ + __ = __

__ + __ = __

www.homerunpress.com

1. <u>Complete</u> each picture. When a shape is symmetrical, each half is a mirror image of the other.

Symmetry

line of symmetry

2. <u>Add.</u>

1	2	3	6	4	8
+ 7	+ 5	+ 3	+ 2	+ 1	+ 1
8					

2	1	3	2	3	6
+ 6	+ 5	+ 5	+ 2	+ 4	+ 1

Subtraction is the opposite of addition. Subtraction means finding the difference between two numbers or taking away from a number. When I give 2 candies to my sister out of 3 candies that I have, how many candies are left?

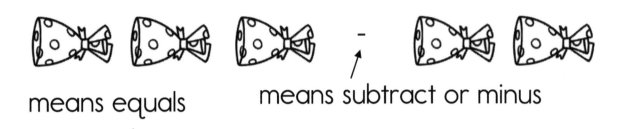

means equals   means subtract or minus

3 - 2 = 1 candy

When I subtract or take away 2 cars from the 4 cars that my brother has, he is left with 2 cars.

He has 4 cars and I take away 2 cars. I can find the total simply by crossing out the 2 cars from the 4 cars. There are 2 cars left. 4 - 2 = 2

1. Read.

I use a number line to find out the answer when I subtract 4 from 7. First, I <u>draw</u> a line and <u>mark</u> it with numbers. I find 7 on the number line.

Start counting at 7.

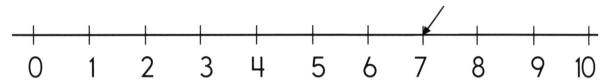

I need to take away 4, so I jump 4 places to the left.

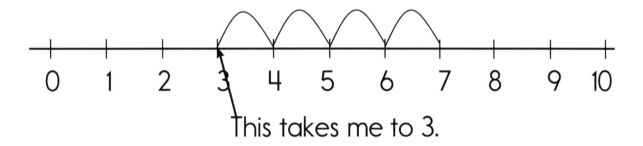

This takes me to 3.

So   7 - 4 = 3

I subtract 40 from 60. First, I find 60. Then, I jump 4 places to the left.

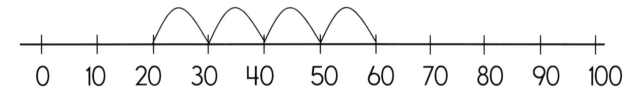

So   60 - 40 = 20

1. <u>Subtract</u>. Use a number line to show the jumps.

9 - 6 = ___

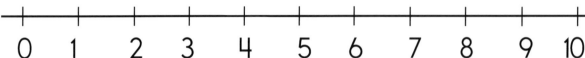

7 - 5 = ___

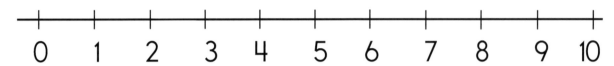

10 - 7 = ___

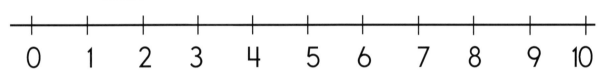

80 - 30 = ___

60 - 50 = ___

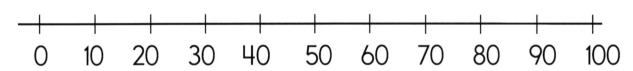

100 - 70 = ___

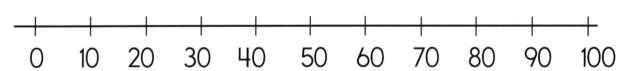

1. Read.

I like to split the numbers I subtract into numbers that are easier to work with. I can show my favorite strategy.

Step 1. Let's subtract 13 from 38.

T O	T O	
3 8	- 1 3	= ___

Step 2. Subtract the tens from 38.

T O	T O	T O
3 8	- 1 0	= 2 8

Step 3. Subtract the ones from the remaining 38.

T O	T O	T O
3 8	- 3	= 3 5

2. Subtract.

5 – 1 = ___   6 – 1 = ___   9 – 1 = ___

8 – 2 = ___   4 – 2 = ___   7 – 2 = ___

3 – 3 = ___   6 – 3 = ___   8 – 3 = ___

7 – 4 = ___   9 – 4 = ___   5 – 4 = ___

I can subtract numbers using column subtraction.
Hint: Write ones under ones. Write tens under tens.

Step 1: Write the digits that have the same place value lined up one above the other.

```
tens  ones
  2     9
- 1     3
-----------
  _     _
```

Step 2: Subtract 3 ones from 9 ones: 9 - 3 = 6

Write 6 in the ones column.

```
tens  ones
  2    (9)
- 1    (3)
-----------
  _    (6)
```

Step 3: Subtract 1 ten from 2 tens:

2 - 1 = 10

I write 1 in the tens column.

```
 tens  ones
( 2 )   9
(-1 )   3
-----------
  1     6
```

1. Subtract.

$$
\begin{array}{cc}
 & 9 \\
- & 3 \\
\hline
 & 6
\end{array}
\qquad
\begin{array}{cc}
 & 5 \\
- & 2 \\
\hline
\end{array}
\qquad
\begin{array}{cc}
 & 8 \\
- & 4 \\
\hline
\end{array}
\qquad
\begin{array}{cc}
 & 7 \\
- & 2 \\
\hline
\end{array}
\qquad
\begin{array}{cc}
 & 4 \\
- & 4 \\
\hline
\end{array}
\qquad
\begin{array}{cc}
 & 8 \\
- & 1 \\
\hline
\end{array}
$$

$$
\begin{array}{cc}
 & 8 \\
- & 2 \\
\hline
\end{array}
\qquad
\begin{array}{cc}
 & 5 \\
- & 5 \\
\hline
\end{array}
\qquad
\begin{array}{cc}
 & 5 \\
- & 2 \\
\hline
\end{array}
\qquad
\begin{array}{cc}
 & 7 \\
- & 6 \\
\hline
\end{array}
\qquad
\begin{array}{cc}
 & 9 \\
- & 1 \\
\hline
\end{array}
\qquad
\begin{array}{cc}
 & 6 \\
- & 4 \\
\hline
\end{array}
$$

2. Complete each pair of number bonds.

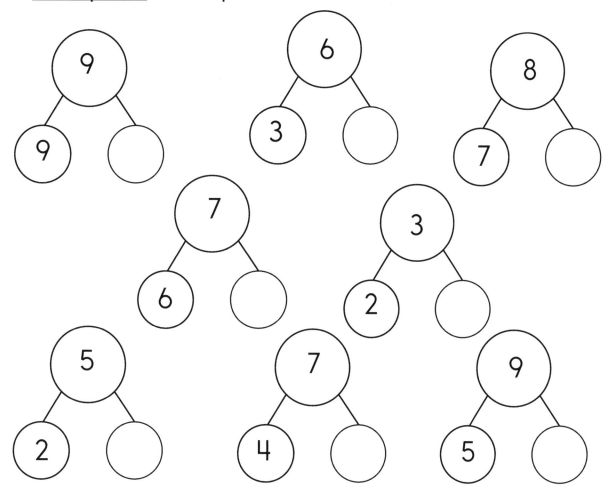

1. <u>Subtract.</u>

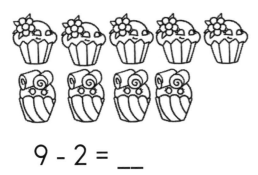

9 - 2 = __

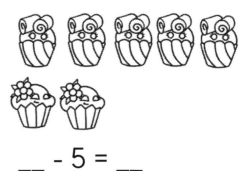

__ - 5 = __

__ - 4 = __

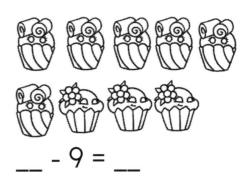

__ - 9 = __

__ - 3 = __

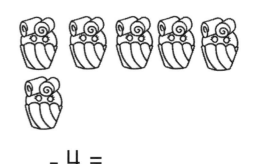

__ - 4 = __

__ - 6 = __

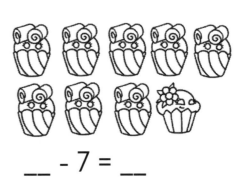

__ - 7 = __

www.homerunpress.com

1. If 5 cupcakes are eaten, <u>how many</u> are left?

10 - 5 = ___

2. If 7 cupcakes are eaten, <u>how many</u> are left?

__ - __ = __

3. If 4 cupcakes are eaten, <u>how many</u> are left?

__ - __ = __

4. If 9 cupcakes are eaten, how many are left?

__ - __ = __

1. <u>Subtract.</u>

__ - 5 = __   __ - 7 = __

__ - 4 = __   __ - 2 = __

2. <u>Subtract.</u>

9	8	6	6	4	8
- 7	- 5	- 3	- 1	- 3	- 4
2					

7	7	8	2	6	6
- 4	- 5	- 3	- 2	- 4	- 5

www.homerunpress.com

1. <u>Color</u> the flowers with the smaller number in each group.

1. Subtract.

10	10	10	10	10	10
- 3	- 5	- 6	- 2	- 4	- 1
7					

17	18	13	15	16	19
- 6	- 5	- 1	- 2	- 4	- 1

2. Complete each pair of number bonds.

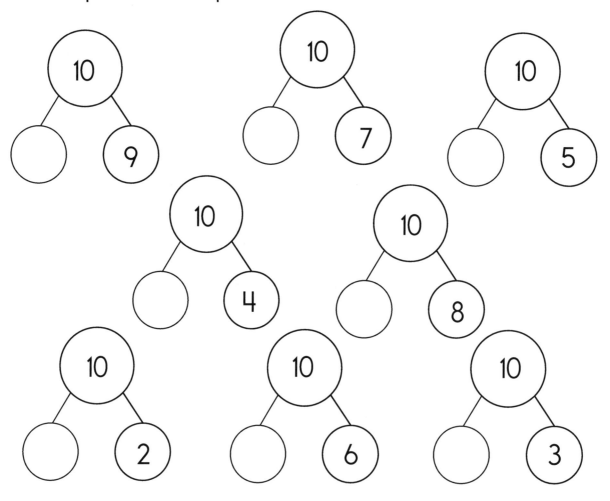

© 2020 Home Run Press, LLC   www.homerunpress.com

1. <u>Read</u>. <u>Write</u> how many tens. <u>Write</u> the number.

Two digit numbers are made up of tens and ones.

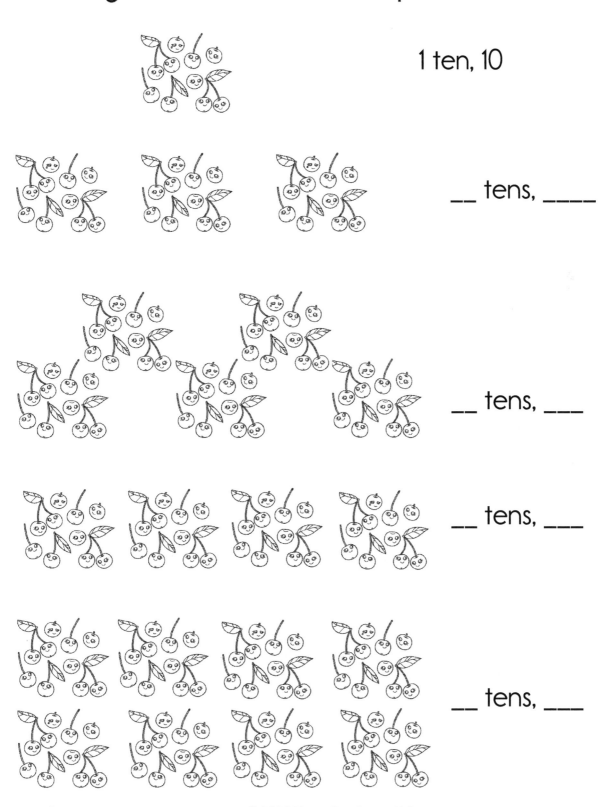

1 ten, 10

__ tens, ____

__ tens, ___

__ tens, ___

__ tens, ___

1. <u>How many</u> are there in each group?

10

1

10 + 1 makes 11

__

_

__ + __ makes __

__

_

__ + __ makes __

__

_

__ + __ makes __

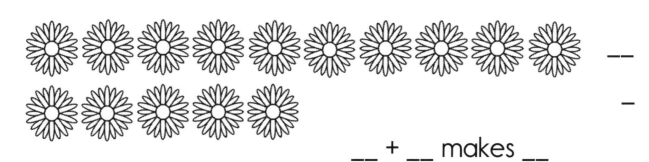

__

_

__ + __ makes __

1. <u>Add.</u>

10	10	10	10	10	10
+ 3	+ 5	+ 6	+ 2	+ 4	+ 1
13					

12	11	13	12	13	16
+ 6	+ 5	+ 5	+ 2	+ 4	+ 1

2. <u>Complete</u> each pair of number bonds.

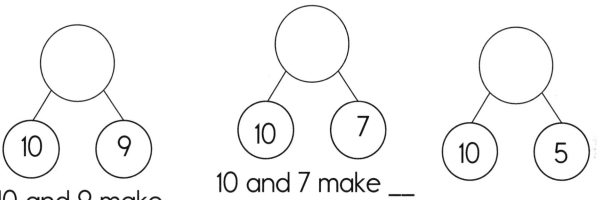

10 and 9 make __

10 and 7 make __

10 and 5 make __

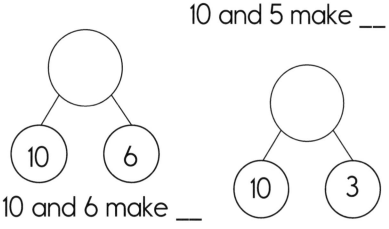

10 and 2 make __

10 and 6 make __

10 and 3 make __

1. Add.

```
   1  1        3  2        2  3        8  1        5  4
+     5     +     6     +     1     +     7     +     2
─────────   ─────────   ─────────   ─────────   ─────────
   1  6        _  _        _  _        _  _        _  _
```

```
   3  1        5  2        9  1        4  5        4  0
+     4     +     2     +     4     +     1     +     8
─────────   ─────────   ─────────   ─────────   ─────────
   _  _        _  _        _  _        _  _        _  _
```

```
   2  5        3  3        6  1        1  2        6  7
+  2  2     +  3  6     +  1  5     +  4  3     +  1  2
─────────   ─────────   ─────────   ─────────   ─────────
   _  _        _  _        _  _        _  _        _  _
```

1. Subtract.

```
   1  9        8  7        4  8        3  5        6  9
-     4     -     2     -     3     -     1     -     3
─────────   ─────────   ─────────   ─────────   ─────────
   1  5        _  _        _  _        _  _        _  _
```

```
   5  9        6  9        7  6        9  4        2  8
-  2  1     -  4  5     -  3  4     -  5  3     -  1  8
─────────   ─────────   ─────────   ─────────   ─────────
   _  _        _  _        _  _        _  _        _  _
```

```
   3  5        8  7        9  8        9  8        7  7
-  1  1     -  3  3     -  3  4     -  5  3     -  1  5
─────────   ─────────   ─────────   ─────────   ─────────
   _  _        _  _        _  _        _  _        _  _
```

1. <u>Circle</u> the missing number from the choice box to make the inequality true.

20 < ___ < 31	a) 18	b) 32	c) 26
9 < ___ < 11	a) 15	b) 10	c) 7
5 < ___ < 15	a) 11	b) 20	c) 4
46 < ___ < 50	a) 38	b) 48	c) 58

2. <u>What</u> is the value of the 3 in each of these numbers? <u>Circle</u> the right answer.

613	a) Hundreds	b) Tens	c) Ones
352	a) Hundreds	b) Tens	c) Ones
134	a) Hundreds	b) Tens	c) Ones
943	a) Hundreds	b) Tens	c) Ones

3. I saw 8 frogs. 5 left. <u>How many frogs</u> stayed?

Answer: _____

1. <u>How many</u> are there in each group?

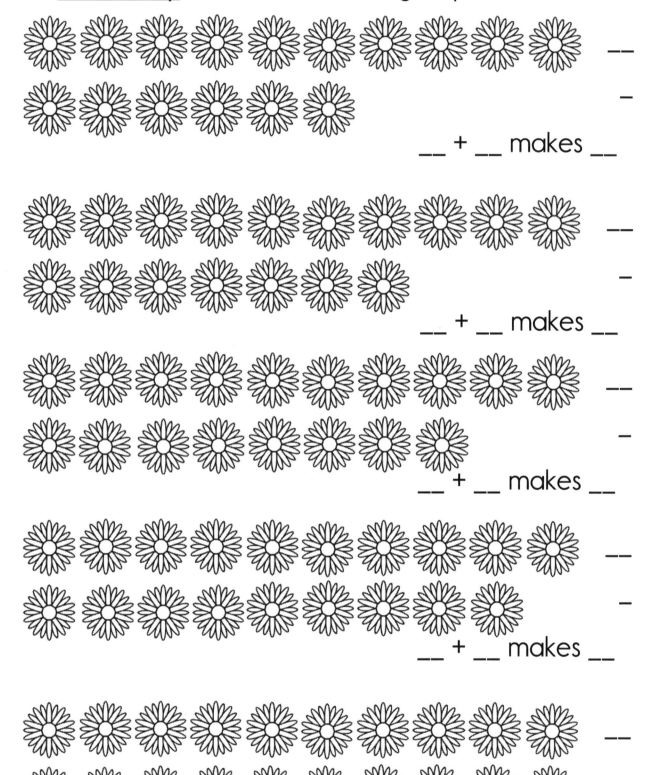

__

–

__ + __ makes __

__

–

__ + __ makes __

__

–

__ + __ makes __

__

–

__ + __ makes __

__

–

__ + __ makes __

www.homerunpress.com

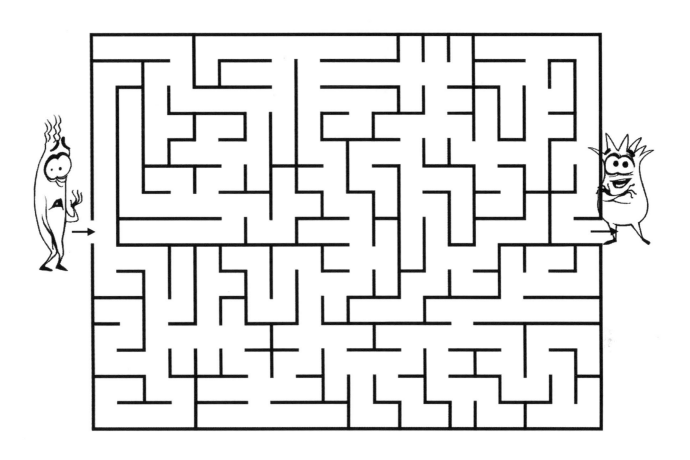

1. <u>Solve</u> the problem and <u>circle</u> ">," "<," or "=:"

1 cupcake   =   10 candies.

3 cupcakes   > / < / =   2 cupcakes + 5 candies

Answer: _____.

1. <u>Add.</u>

```
   20        20        30        30        30        40
 +  3      +  5      +  6      +  2      +  4      +  1
 ─────     ─────     ─────     ─────     ─────     ─────
   23
```

```
   42        51        53        62        63        66
 +  6      +  5      +  5      +  2      +  4      +  1
 ─────     ─────     ─────     ─────     ─────     ─────
```

2. <u>Complete</u> each pair of number bonds.

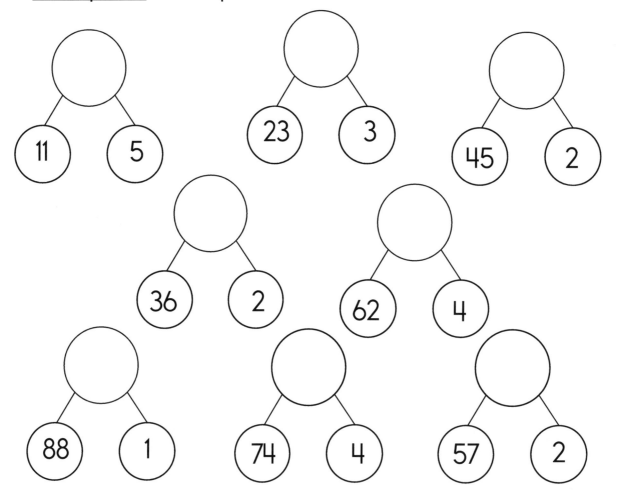

© 2020 Home Run Press, LLC   www.homerunpress.com

1. The blocks in each tower tell you how many hundreds, tens, and ones in each number. <u>Write and</u> put the numbers in order from the least to the

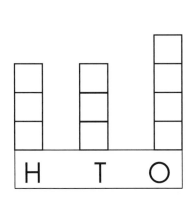 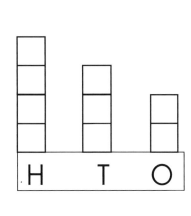 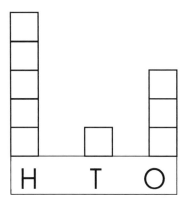

_____   _____   _____

2. <u>Write</u> the number for these.

2 tens + 5 ones = 25   1 ten + 9 ones = _____

3 tens + 0 ones = _____   5 tens + 6 ones = _____

4 tens + 9 ones = _____   6 tens + 2 ones = _____

7 tens + 0 ones = _____   8 tens + 5 ones = _____

3. <u>What</u> is the value of the digit 7 in the numbers below?

37   71   17   7   79

1. <u>Add.</u>

11	14	15	17	13	12
+ 12	+ 14	+ 21	+ 31	+ 14	+ 16
23					

22	41	33	82	73	56
+ 12	+ 41	+ 25	+ 14	+ 26	+ 13

2. <u>Complete</u> each pair of number bonds.

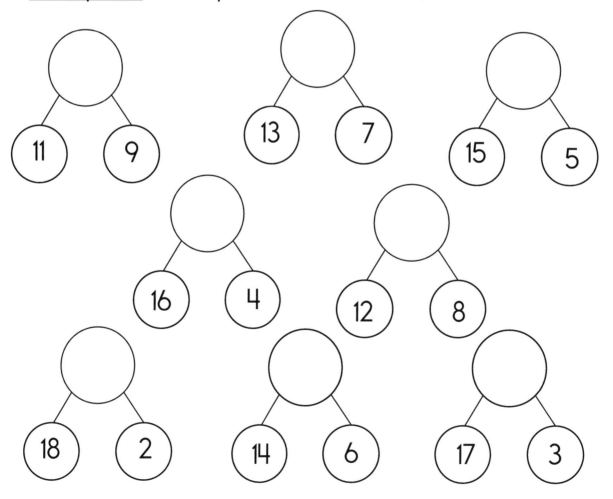

www.homerunpress.com

1. Subtract.

25	28	37	33	37	42
- 3	- 5	- 6	- 2	- 4	- 1
22					

47	56	58	67	69	66
- 6	- 5	- 5	- 2	- 4	- 5

2. Complete each pair of number bonds.

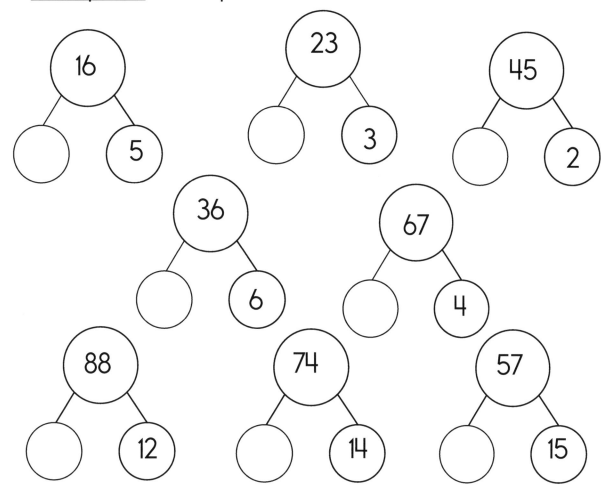

1. Subtract.

34	28	45	37	57	47
- 12	- 14	- 21	- 31	- 14	- 16
22					

22	64	57	39	48	33
- 12	- 41	- 25	- 14	- 26	- 11

2. Complete each pair of number bonds.

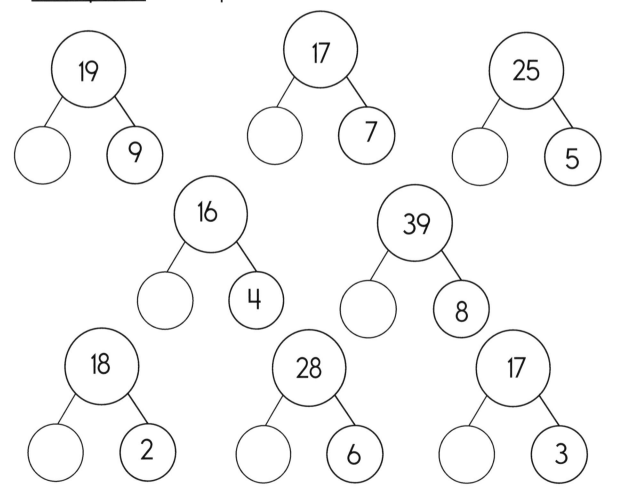

www.homerunpress.com

1. <u>How many more</u> do I need to add to the second group to make each group the same?

 = 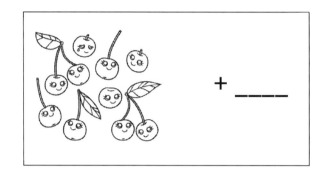 + _____

 =  + _____

 = 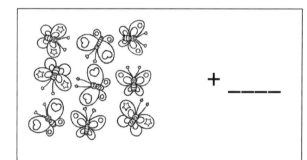 + _____

2. <u>Circle</u> all the combinations that equal 5.

10 – 5   16 – 11   8 – 2

   14 – 12   21 – 10   29 – 24

38 – 28   17 – 12   46 - 40

1. If 20 pencils are broken, <u>how many</u> are left?

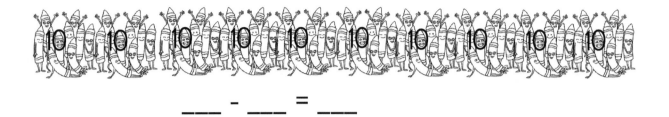

___ - ___ = ___

2. If 30 pencils are broken, <u>how many</u> are left?

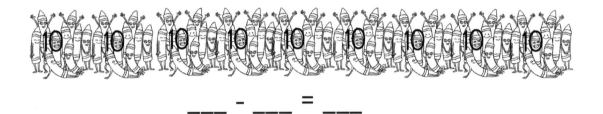

___ - ___ = ___

3. If 50 pencils are broken, <u>how many</u> are left?

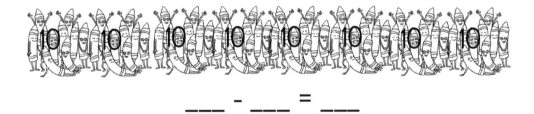

___ - ___ = ___

4. If 40 pencils are broken, <u>how many</u> are left?

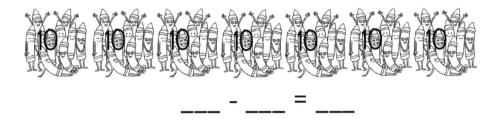

___ - ___ = ___

1. Venn Diagram: helps you sort things according to their different features.

I have many cards. 3 of them are dragon-type cards. 5 of them are flying-type cards. 2 of them are both dragon- and flying-type cards. 2 of them are neither dragon- nor flying-type cards. How many cards are there? Fill in the diagram.

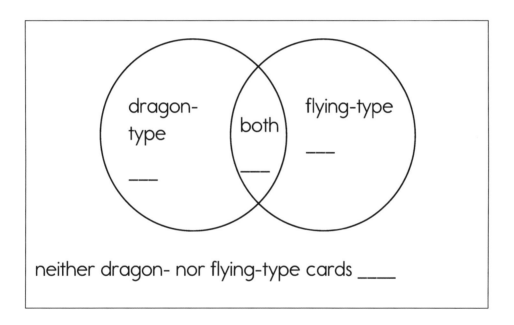

dragon-type

both

flying-type

___

___

___

neither dragon- nor flying-type cards ____

Answer: _____

2. Circle the right answer.

a) 1
b) 2
c) 3

__ 2 + 3 6 = 58

1. <u>Solve</u> the problem and <u>write</u> the missing number:

1 sunflower   =   3 tulips

3 sunflowers   =   __   tulips

Answer: _____.

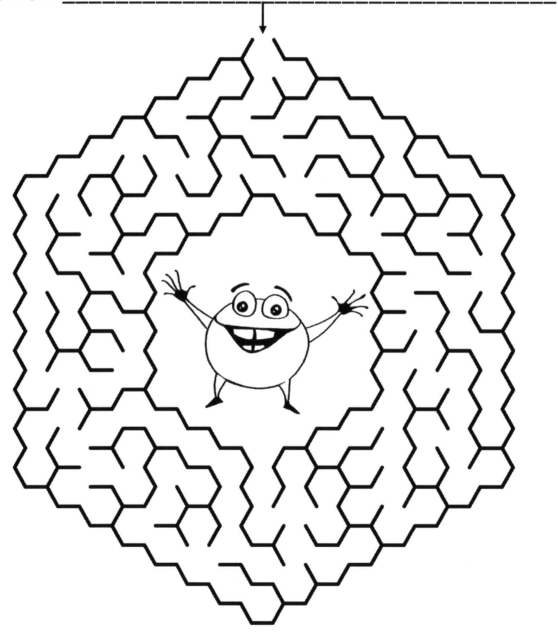

1. Read. Write how many tens. Write the number.

Three digit numbers are made up of hundreds, tens, and ones.

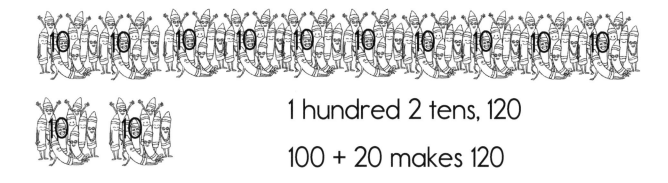

1 hundred 2 tens, 120

100 + 20 makes 120

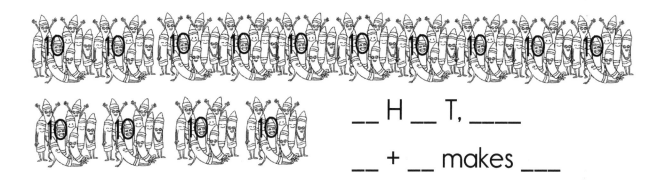

__ H __ T, ____

__ + __ makes ___

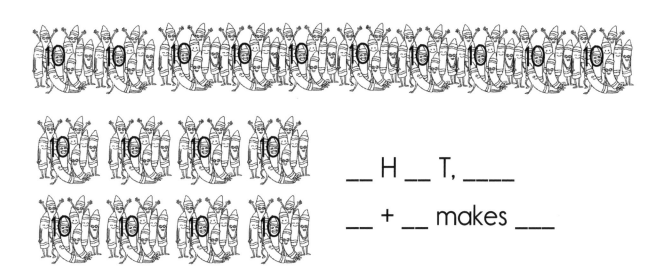

__ H __ T, ____

__ + __ makes ___

1. <u>Add.</u>

100	100	100	100	100	100
+ 25	+ 15	+ 46	+ 12	+ 47	+ 71
125					

120	110	130	120	130	160
+ 6	+ 5	+ 5	+ 2	+ 4	+ 1

2. Complete each pair of number bonds.

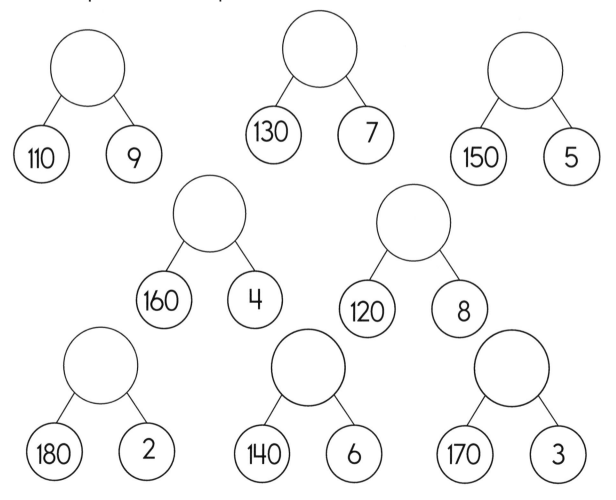

www.homerunpress.com

1. I am building a solid slab of rocks: the two rocks next to each other are added to get the number up above. <u>Fill in</u> the missing numbers.

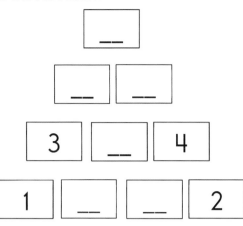

2. <u>Write</u> the numbers in order, from the smallest to the largest.

87, 9, 36, 14, 20, 71, 59, 23, 18, 65

_____

3. Complements to 20. <u>Circle</u> the missing numbers from the choice box to make the equations true.

8 + ___ = 20   a) 8   b) 2   c) 12

11 + ___ = 20   a) 10   b) 8   c) 9

5 + ___ = 20   a) 5   b) 15   c) 10

7 + ___ = 20   a) 13   b) 23   c) 3

2. <u>Circle</u> the right answer.

                                              a)   5
                                              b)   9
__ 6 + 2 3 = 99   c)   7

1. Read.

Place value is the amount a digit is worth in a number.

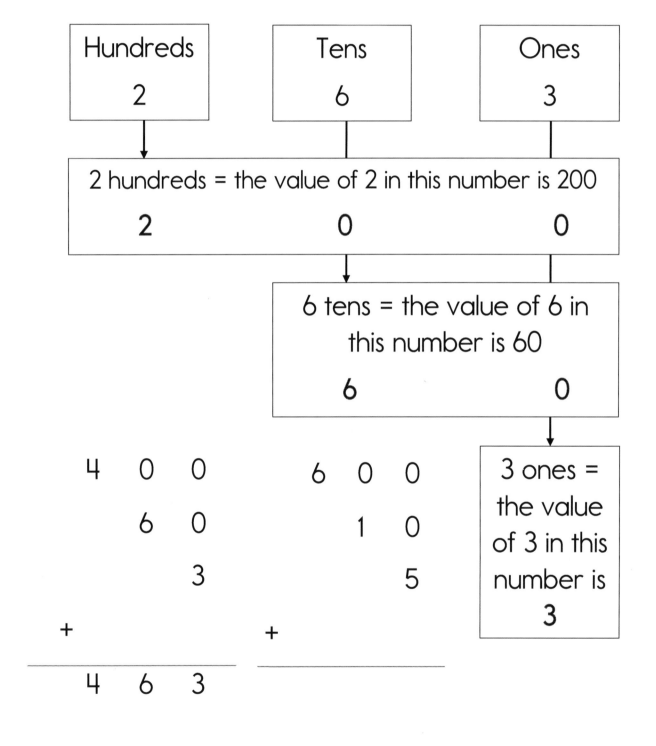

1. <u>What</u> is the value of the 4 in each of these numbers? <u>Circle</u> the right answer.

416   a) Hundreds   b) Tens   c) Ones

914   a) Hundreds   b) Tens   c) Ones

416   a) Hundreds   b) Tens   c) Ones

245   a) Hundreds   b) Tens   c) Ones

624   a) Hundreds   b) Tens   c) Ones

2. <u>Complete</u> each addition number sentence with tens and ones. The first one is done for you.

45 = 40 + 5   61 = __ + __   19 = __ + __

30 = __ + __   74 = __ + __   37 = __ + __

52 = __ + __   65 = __ + __   58 = __ + __

98 = __ + __   87 = __ + __   60 = __ + __

1. Write the missing numbers.

$50 + \_\_\_ = 53$

$20 + \_\_\_ = 25$

$90 + \_\_\_ = 96$

$60 + \_\_\_ = 62$

$10 + \_\_\_ = 19$

$40 + \_\_\_ = 48$

$\_\_\_ + \_\_\_ + 7 = 807$

$\_\_\_ + \_\_\_ + 8 = 328$

$\_\_\_ + \_\_\_ + 2 = 972$

$\_\_\_ + \_\_\_ + 0 = 430$

$\_\_\_ + \_\_\_ + 4 = 244$

$\_\_\_ + \_\_\_ + 5 = 595$

2. Circle the correct answer.

I have a series of numbers: 0, 2, 4, 6, ___. What is the next number?

a) 10   b) 8   c) 12   d) 7

www.homerunpress.com

1. Circle the correct answer.

I have a series of numbers: 1, 2, 4, ___. What is the next number?

   a) 7   b) 8   c) 9   d) 10

2. Color the 2nd frog green. Color the 3rd frog brown. Color the 5th frog black. Color the 6th frog grey.

3. Write the missing numbers.

___ + ___ + ___ = 468   ___ + ___ + ___ = 509

___ + ___ + ___ = 135   ___ + ___ + ___ = 683

___ + ___ + ___ = 245   ___ + ___ + ___ = 248

___ + ___ + ___ = 621   ___ + ___ + ___ = 791

1. Subtract.

146	138	157	134	159	198
- 25	- 15	- 46	- 12	- 47	- 71
121					

127	119	137	126	138	165
- 14	- 15	- 15	- 24	- 31	- 41

2. Complete each pair of number bonds.

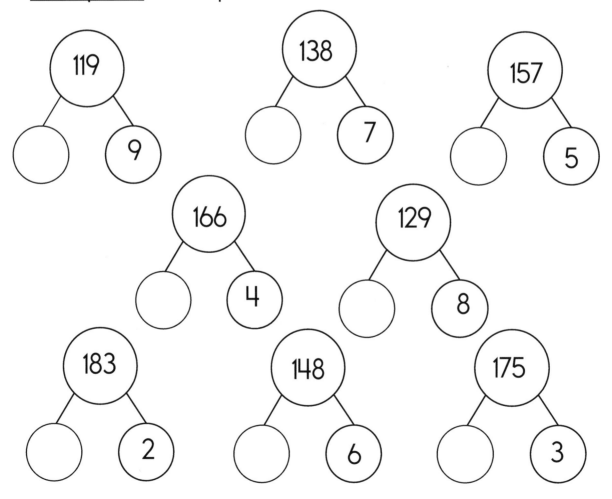

www.homerunpress.com

1. Complements to 100. <u>Circle</u> the missing numbers from the choice box to make the equations true.

70 + ___ = 100   a) 35   b) 30   c) 10

50 + ___ = 100   a) 40   b) 100   c) 50

90 + ___ = 100   a) 10   b) 20   c) 1

2. <u>Subtract</u>.

$$\begin{array}{r} 3\ 5 \\ -\ \ \ 5 \\ \hline 3\ 0 \end{array} \qquad \begin{array}{r} 7\ 2 \\ -\ \ \ 2 \\ \hline \end{array} \qquad \begin{array}{r} 5\ 4 \\ -\ \ \ 4 \\ \hline \end{array} \qquad \begin{array}{r} 6\ 8 \\ -\ \ \ 8 \\ \hline \end{array} \qquad \begin{array}{r} 8\ 7 \\ -\ \ \ 7 \\ \hline \end{array}$$

$$\begin{array}{r} 3\ 5 \\ -\ \ \ 4 \\ \hline \end{array} \qquad \begin{array}{r} 7\ 7 \\ -\ \ \ 2 \\ \hline \end{array} \qquad \begin{array}{r} 6\ 4 \\ -\ \ \ 3 \\ \hline \end{array} \qquad \begin{array}{r} 8\ 2 \\ -\ \ \ 1 \\ \hline \end{array} \qquad \begin{array}{r} 4\ 9 \\ -\ \ \ 5 \\ \hline \end{array}$$

$$\begin{array}{r} 6\ 5 \\ -\ 4\ 0 \\ \hline \end{array} \qquad \begin{array}{r} 9\ 3 \\ -\ 2\ 0 \\ \hline \end{array} \qquad \begin{array}{r} 5\ 7 \\ -\ 1\ 0 \\ \hline \end{array} \qquad \begin{array}{r} 3\ 9 \\ -\ 2\ 0 \\ \hline \end{array} \qquad \begin{array}{r} 8\ 1 \\ -\ 7\ 0 \\ \hline \end{array}$$

3. The sum of the two 2-digit numbers is 50. Their difference is 30. What are these 2-digit numbers?

Answer: __ __ and __ __.

__ __ + __ __ = __ __   __ __ - __ __ = __ __

1. <u>Read</u>.

I often need to know if a number is the same as, smaller than, or larger than another <u>number</u>. My teacher calls this comparing numbers. Look at these candies. There are six candies in each row. My teacher says that the number in one row is **equal to** the number in the second row. **6 = 6**

My sister has six candies in the top row and three candies in the bottom row. She says the number in the top row is **greater than** the number in the bottom row. **6 > 3. 6 is greater than 3.**

1. <u>Read.</u>

My brother has five candies in the top row and six candies in the bottom row. He says the number in the top row is **less than** the number in the second row.   5 < 6.   5 is less than 6.

2.   <u>Circle</u> the missing number from the choice box to make the inequality true.

5 < ___< 9

a) 4   b) 0   c) 6

9 < ___< 11

a) 7   b) 1   c) 10

15 < ___< 25

a) 11   b) 20   c) 35

76 < ___< 90

a) 89   b) 71   c) 100

1. <u>Read.</u>

When I round, I change a number to another number that is almost the same in value, but it is easier to work with.

For digits 0, 1, 2, and 4, we round down

For digits 5, 6, 7, 8, and 9, we round up

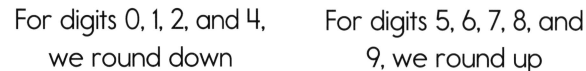

Look at 5<u>2</u>. We look at the ones digit. It's 2.

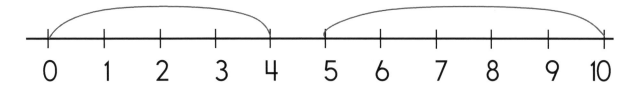

We round down to 50.

Now, look at 5<u>8</u>. The ones digit is 8.

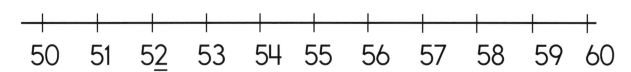

So we round UP to 60.

 www.homerunpress.com

1. <u>Which</u> is more? <u>Compare</u> the numbers using ">," "<," or "=."

3 ones _____ 2 tens   10 ones _____ 1 ten

3 tens _____ 8 ones   2 tens _____ 1 ten

4 tens 2 ones _____ 41 ones

1 ten 2 ones _____ 1 ten 20 ones

2 tens and 3 ones _____ 3 tens and 2 ones

2. <u>Round</u> each number to the nearest 10. <u>Look</u> at the next digit to the right. If it is 0, 1, 2, 3, or 4, then ROUND DOWN, if it is 5, 6, 7, 8, 9, then ROUND UP.

<u>6</u>   _____   <u>8</u> _____   <u>5</u> _____

1<u>3</u>   _____   1<u>9</u> _____   1<u>6</u> _____

1<u>1</u>   _____   1<u>7</u>   _____   1<u>9</u> _____

2<u>2</u> _____   2<u>5</u> _____   2<u>3</u> _____

2<u>8</u> _____   2<u>4</u> _____   2<u>9</u> _____

1. <u>Which</u> is more? <u>Write</u> the missing numbers to make the comparison true.

12 ones > ___ ones   7 ones < ___ ten

1 ten = ___ ones   2 tens = ___ ones

5 ones > ___ ones   12 ones < ___ tens

15 ones < ___ ten 6 ones

1 ten 8 ones > 1 ten ___ ones

3 tens and 3 ones = ___ tens and 13 ones

2. <u>Round</u> each number to the nearest 100. <u>Look</u> at the next digit to the right. If it is 0, 1, 2, 3, or 4, then ROUND DOWN, if it is 5, 6, 7, 8, 9, then ROUND UP.

1<u>5</u>3 _____   2<u>0</u>8 _____   7<u>1</u>5 _____

9<u>1</u>3 _____   2<u>5</u>9 _____   2<u>4</u>6 _____

3<u>7</u>1 _____   5<u>5</u>7 _____   4<u>6</u>9 _____

6<u>2</u>2 _____   4<u>8</u>5 _____   9<u>2</u>3 _____

          www.homerunpress.com

1. Read.

Even numbers are made of pairs.

An odd number is always 1 more or 1 less than an even number.

Even numbers end with a digit of 0, 2, 4, 6, 8.

Odd numbers end with a digit of 1, 3, 5, 7, 9.

2. Underline the even numbers.

1, 2, 3, 4, 5, 6, 7, 8, 9, 10, 11, 12, 13, 14, 15, 16

3. Circle the odd numbers.

15, 16, 17, 18, 19, 20, 21, 22, 23, 24, 25, 26

1. <u>Read.</u>

I have tons of candies. I need to estimate because it would take too long to count the exact number. I count 5 candies in the bottom row. There are 4 rows, so I can say there are about 5 + 5 + 5 + 5 candies, which is 20 candies.

I often don't need to count the candies exactly. If I have two bags of candies that cost the same, I will get the bag with more candies.

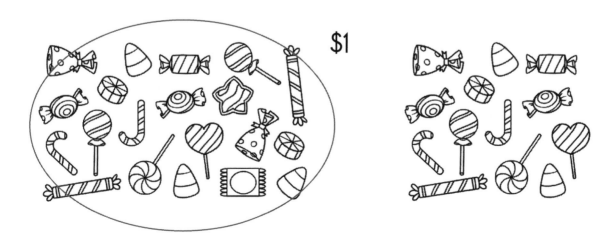

www.homerunpress.com

1. I need money to buy objects. <u>Write</u> the missing numbers.

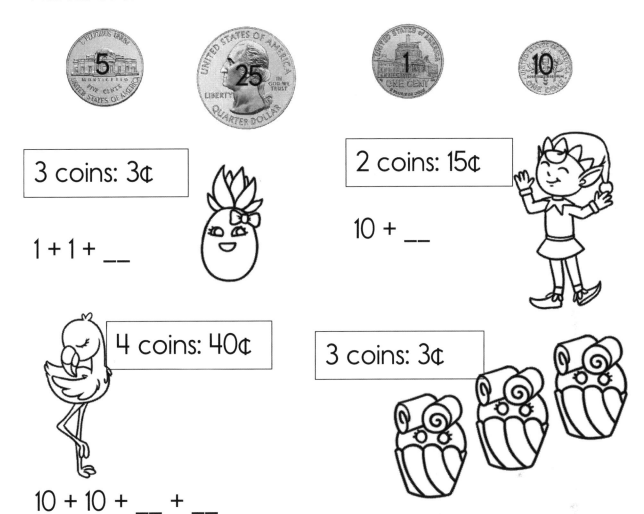

3 coins: 3¢

1 + 1 + __

2 coins: 15¢

10 + __

4 coins: 40¢

10 + 10 + __ + __

3 coins: 3¢

1 + __ + __

2 coins: 10¢

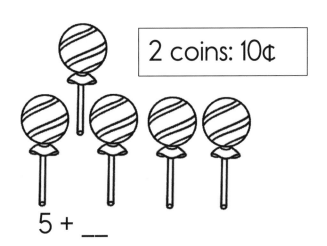

5 + __

I bought 5 lollipops for 10 cents. How much did one lollipop cost?

_____

1. I need money to buy objects. <u>Write</u> the missing numbers.

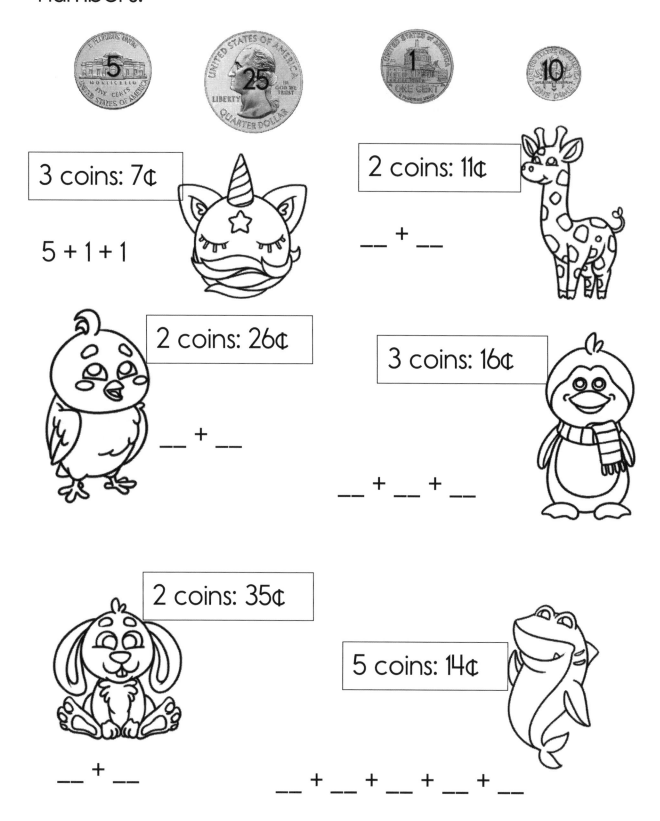

3 coins: 7¢

5 + 1 + 1

2 coins: 11¢

__ + __

2 coins: 26¢

__ + __

3 coins: 16¢

__ + __ + __

2 coins: 35¢

__ + __

5 coins: 14¢

__ + __ + __ + __ + __

www.homerunpress.com

1. I need money to buy objects. <u>Write</u> the missing numbers.

3 coins: 15¢

__ + __ + __

2 coins: 15¢

__ + __

2 coins: 20¢

__ + __

4 coins: 17¢

__ + __ + __ + __

2 coins: 30¢

__ + __

5 coins: 17¢

__ + __ + __ + __ + __

1. <u>Write</u> the missing numbers to make the scales balance.

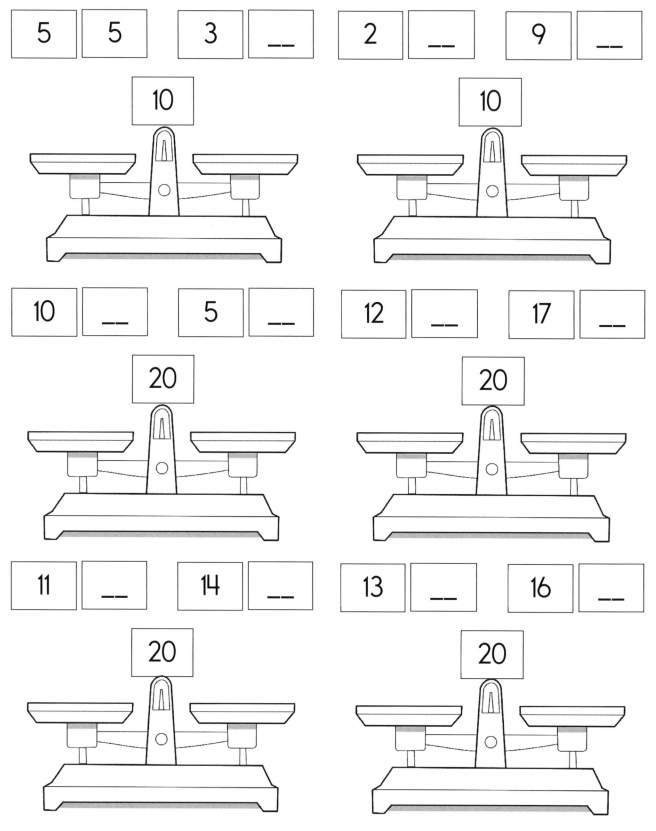

1. <u>Write</u> the missing numbers to make the scales balance.

15	5

18	__

20	__

19	__

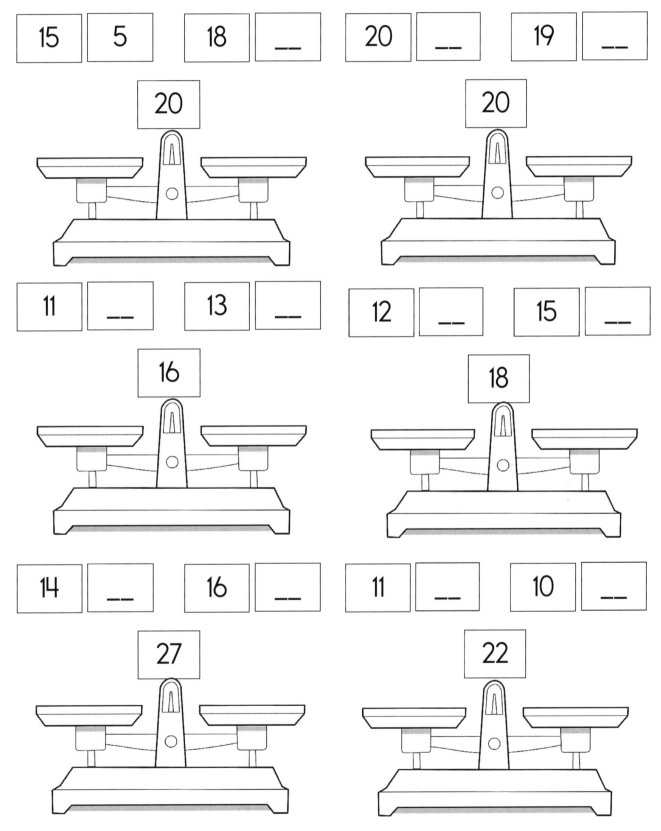

20   20

11	__

13	__

12	__

15	__

16   18

14	__

16	__

11	__

10	__

27   22

1. My Grandma baked 2 pumpkin pies. I ate a half of the total amount of pies. How many pie(s) are left?

Circle your answer:

0 ① 2 3 4 5

A half is one of two equal parts of one whole. If two pies are one whole, I could eat 1 pie which is a half. Another half is left. So, I circle 1.

I found 4 shells. My sister broke a half of the shells. Color these shells red. How many shells are left?

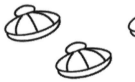

Circle your answer:

0 1 2 3 4 5

I got 6 cupcakes. I ate a half of them. Color the cupcakes I ate. How many cupcakes are left?

Circle your answer:

0 1 2 3 4 5

www.homerunpress.com

1. I got 8 candies. I ate a half of the candies. Color them red. How many candies are left?

Circle your answer:

0   1   2   3   4   5

I found 10 flowers. A half of the flowers were blooming. How many flowers were not blooming?

Circle your answer:

0   1   2   3   4   5

The pumpkin weighed 2 pounds. We ate a half of it. How many pounds are left?

Circle your answer:

0   1   2   3   4   5

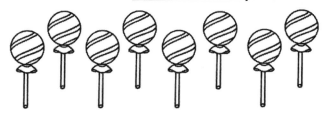

My birthday cake weighed 10 pounds! My friends ate a half of the cake. How many pounds are left?

Circle your answer: 0   1   2   3   4   5

1. <u>What number</u> am I?

Half of me is 1 and double me is 4.

One whole:

A half of two equal parts of one whole is one:

Double means take twice as much or as many:

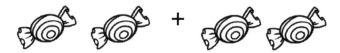

Half of me is 2 and double me is 8.

I am _____.

Half of me is 5 and double me is 20.

I am _____.

Half of me is 10 and double me is 40.

I am _____.

Half of me is 50 and double me is 200.

I am _____.

www.homerunpress.com

1. When you share equally between two elves, both sets of sweets and fruits have the same amount. <u>Count how many</u> for earch elf?

Each elf must have the same amount.

_____	_____	_____	_____

1. I am building a solid slab of rocks: the two rocks next to each other are added to get the number up above. <u>Fill in</u> the missing numbers.

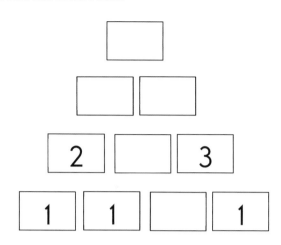

2. <u>Write</u> the numbers in order, from the smallest to the largest.

7, 9, 3, 1, 2, 11, 5, 12, 8, 4, 0, 10, 6

_____

3. Complements to 10. <u>Circle</u> the missing numbers from the choice box to make the equations true.

11 + ___ = 20   a) 8   b) 7   c) 9

5 + ___ = 20   a) 10   b) 5   c) 15

7 + ___ = 20   a) 13   b) 1   c) 14

14 + ___ = 20   a) 4   b) 16   c) 6

www.homerunpress.com

1. Circle the right answer.

d) 25

e) 30

f) 35

___ + 25 = 55

2. Complements to 100. Circle the missing numbers from the choice box to make the equations true.

50 + ___ = 100   a) 35   b) 50   c) 60

90 + ___ = 100   a) 10   b) 100   c) 20

20 + ___ = 100   a) 60   b) 20   c) 80

3. Circle the right answer:

I have a series of numbers: 3, 6, 9, 12, __

What is the next number?

  A) 13   B) 15   C) 18   D) 14

4. I have some numbers and signs: 1, 5, 3, +, -.

Write the equation that equals one of the answer choices. _____

  A) 10   B) 7   C) 4   D) 9

1. <u>Which</u> is more? <u>Compare</u> the numbers using ">," "<," or "=."

13 ones _____ 2 tens   9 ones _____ 1 ten

3 tens _____ 30 ones   14 tens _____ 1 hundred

4 hundreds _____ 41 tens   2 hundreds _____ 199 ones

3 tens and 5 ones _____ 5 tens and 3 ones

2. <u>Round</u> each number to the nearest 10. <u>Look</u> at the next digit to the right. If it is 0, 1, 2, 3, or 4 then ROUND DOWN, if it is 5, 6, 7, 8, 9 then, ROUND UP.

23 _____   44 _____   26 _____   51 _____

87 _____   39 _____   72 _____   65 _____

3. <u>What number</u> am I?

Half of me is 20 and double me is 80. _____

Half of me is 10 and double me is 40. _____

Half of me is 25 and double me is 100. _____

Half of me is 11 and double me is 44. _____

1. <u>Solve</u> the problem:

1047: the sum of the ones and hundreds is _____.

A) 5
B) 8
C) 7

2. I start at 0 and count on in twos. Will I say 11?

_____

Why? _____

I start at 0 and count on in twos. Will I say 16?

_____

Why? _____

3. <u>Find</u> the value.

853:

The sum of the ones and tens is _____.

The difference between the hundreds and tens is _____.

The difference between the hundreds and ones is _____.

1. <u>Solve</u> the problems:

I had 8 candies. I gave 4 of them to my sister. <u>How many candies</u> has I left? _____

There are 10 kids at a playground. I counted 7 boys. <u>How many girls</u> are there?

_____

My brother bought 11 chocolate cupcakes and 5 vanilla cupcakes. <u>How many cupcakes</u> did he buy in all?

_____

I found 6 easter eggs. My sister found 4 more Easter eggs than I did. My brother found 7 less Easter eggs than my sister. <u>How many Easter eggs</u> did my brother find?

_____

www.homerunpress.com

1. <u>Solve</u> the problems:

I had 15 cupcakes. I ate some cupcakes, and I had 12 cupcakes left. <u>How many cupcakes</u> did I eat?

_____

My brother has 18 trucks and race cars. 3 of them are trucks. <u>How many race cars</u> does he have?

_____

My sister saw 11 butterflies. My brother saw 4 butterflies more than my sister. <u>How many butterflies</u> did they see altogether?

_____

1. <u>Solve</u> the problems:

I had some cupcakes. I ate 3 cupcakes and I gave 4 cupcakes to my friend. I have 11 cupcakes left. <u>How many cupcakes</u> did I have at first?

_____

I had 5 yellow balloons and 5 more red balloons than yellow balloons. My friend had 8 more ballons than I had . <u>How many balloons</u> did my friend have?

_____

There are 6 oranges in a basket. My mother puts 10 small pears and 3 bananas into the basket. <u>How many fruits</u> are there in the basket altogether?

_____

Hint: Write ones under ones. Carry over when the sum is 10 or more. Find out how many more you need to add to a greater number to get a ten.

5 + 8

2   3

Decompose a smaller number. It's 5.

5 is 2+3 and I need to add two more to 8 to get 10.

ones
5
+   8

In columns I add one more row to write the numbers that were carried over, right?

tens ones

–

5
+   8

– –

Step 1: I need one more row above 5.

Step 2: First, I add ones: 5+8=13.

tens ones
1

5
8
–   3

Write the 3 in one's place. Carry 1 ten with the tens. Write the 1 in ten's place.

tens ones
1

5
+   8
1   3

Step 3: Add tens: 1+0+0=1. Rewrite the 1 in ten's place.

1. Add.

$$\begin{array}{r} {}^{-}9 \\ +\ 7 \\ \hline \end{array} \qquad \begin{array}{r} {}^{-}9 \\ +\ 6 \\ \hline \end{array} \qquad \begin{array}{r} {}^{-}9 \\ +\ 4 \\ \hline \end{array} \qquad \begin{array}{r} {}^{-}9 \\ +\ 5 \\ \hline \end{array} \qquad \begin{array}{r} {}^{-}9 \\ +\ 8 \\ \hline \end{array}$$

$$\begin{array}{r} {}^{-}9 \\ +\ 2 \\ \hline \end{array} \qquad \begin{array}{r} {}^{-}9 \\ +\ 1 \\ \hline \end{array} \qquad \begin{array}{r} {}^{-}9 \\ +\ 3 \\ \hline \end{array} \qquad \begin{array}{r} {}^{-}8 \\ +\ 6 \\ \hline \end{array} \qquad \begin{array}{r} {}^{-}8 \\ +\ 8 \\ \hline \end{array}$$

$$\begin{array}{r} {}^{-}8 \\ +\ 7 \\ \hline \end{array} \qquad \begin{array}{r} {}^{-}8 \\ +\ 9 \\ \hline \end{array} \qquad \begin{array}{r} {}^{-}8 \\ +\ 4 \\ \hline \end{array} \qquad \begin{array}{r} {}^{-}8 \\ +\ 2 \\ \hline \end{array} \qquad \begin{array}{r} {}^{-}8 \\ +\ 5 \\ \hline \end{array}$$

www.homerunpress.com

1. <u>Add.</u>

$$\begin{array}{r} \bar{8} \\ + \ 3 \\ \hline \end{array} \qquad \begin{array}{r} \bar{7} \\ + \ 6 \\ \hline \end{array} \qquad \begin{array}{r} \bar{7} \\ + \ 4 \\ \hline \end{array} \qquad \begin{array}{r} \bar{7} \\ + \ 5 \\ \hline \end{array} \qquad \begin{array}{r} \bar{7} \\ + \ 8 \\ \hline \end{array}$$

$$\begin{array}{r} \bar{7} \\ + \ 3 \\ \hline \end{array} \qquad \begin{array}{r} \bar{7} \\ + \ 7 \\ \hline \end{array} \qquad \begin{array}{r} \bar{7} \\ + \ 9 \\ \hline \end{array} \qquad \begin{array}{r} \bar{6} \\ + \ 6 \\ \hline \end{array} \qquad \begin{array}{r} \bar{6} \\ + \ 8 \\ \hline \end{array}$$

$$\begin{array}{r} \bar{6} \\ + \ 7 \\ \hline \end{array} \qquad \begin{array}{r} \bar{6} \\ + \ 9 \\ \hline \end{array} \qquad \begin{array}{r} \bar{6} \\ + \ 4 \\ \hline \end{array} \qquad \begin{array}{r} \bar{6} \\ + \ 5 \\ \hline \end{array} \qquad \begin{array}{r} \bar{5} \\ + \ 5 \\ \hline \end{array}$$

2.

Circle the right answer:

I have a series of numbers:
0, 2, 2, 4, 6, __

What is the next number?

A 6   C 10

B 12   D 9

1. <u>Add.</u>

$$\begin{array}{r} 5 \\ + \ 9 \\ \hline \end{array} \qquad \begin{array}{r} 5 \\ + \ 6 \\ \hline \end{array} \qquad \begin{array}{r} 5 \\ + \ 7 \\ \hline \end{array} \qquad \begin{array}{r} 5 \\ + \ 8 \\ \hline \end{array} \qquad \begin{array}{r} 4 \\ + \ 8 \\ \hline \end{array}$$

$$\begin{array}{r} 4 \\ + \ 6 \\ \hline \end{array} \qquad \begin{array}{r} 4 \\ + \ 7 \\ \hline \end{array} \qquad \begin{array}{r} 4 \\ + \ 9 \\ \hline \end{array} \qquad \begin{array}{r} 3 \\ + \ 7 \\ \hline \end{array} \qquad \begin{array}{r} 3 \\ + \ 8 \\ \hline \end{array}$$

$$\begin{array}{r} 3 \\ + \ 9 \\ \hline \end{array} \qquad \begin{array}{r} 2 \\ + \ 9 \\ \hline \end{array} \qquad \begin{array}{r} 8 \\ + \ 9 \\ \hline \end{array} \qquad \begin{array}{r} 1 \\ + \ 9 \\ \hline \end{array} \qquad \begin{array}{r} 5 \\ + \ 5 \\ \hline \end{array}$$

2.

Circle the right answer:

I have a series of numbers:
19, 12, 7, 4, __

What is the next number?

A   5   C   2
B   3   D   1

www.homerunpress.com

1.

Circle the right answer:

I have a series of numbers:

3, 5, 9, __

What is the next number?

A   11   C   15

B   13   D   9

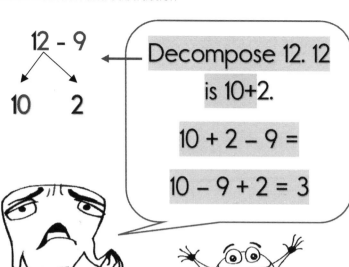

Hint: Write the smaller number under the larger number; Ones under ones, tens under tens; Subtract ones, then tens.

12 - 9

10   2

Decompose 12. 12 is 10+2.

10 + 2 – 9 =

10 – 9 + 2 = 3

tens ones
  1   2
-       9
_____

tens ones –   –
  ‾   ‾
  1   2
-       9
_____
  ‾   ‾

In the columns we subtract 9 ones out of 2 ones.

Step 1: If I subtract in columns, I need one more row above 12.

Step 2: Borrow 1 ten out of the tens since 12=10+2.

Write 12 above 2 in one's place since 10 ones and 2 ones are 12.

tens ones
        12
  1   2
-       9
_____

–   –

Step 3: leave 0 above 1 in ten's place. Cross out 1 and 2 to avoid mistakes. →

Step 3: Subtract 9 ones from 12 ones: 12-9=3. Hint: Do NOT rewrite 0 in ten's place.

tens ones
  0   12
  1̶   2̶
-       9
_____
        3

     www.homerunpress.com

1. <u>Subtract.</u>

$$\begin{array}{r} 0\ \ 14 \\ \cancel{1}\ \ 4 \\ -\quad 9 \\ \hline \end{array}$$

$$\begin{array}{r} \overline{1}\ \ \overline{5} \\ -\quad 8 \\ \hline \end{array}$$

$$\begin{array}{r} \overline{1}\ \ \overline{2} \\ -\quad 6 \\ \hline \end{array}$$

$$\begin{array}{r} \overline{1}\ \ \overline{5} \\ -\quad 7 \\ \hline \end{array}$$

$$\begin{array}{r} \overline{1}\ \ \overline{6} \\ -\quad 8 \\ \hline \end{array}$$

$$\begin{array}{r} \overline{1}\ \ \overline{4} \\ -\quad 7 \\ \hline \end{array}$$

$$\begin{array}{r} \overline{1}\ \ \overline{7} \\ -\quad 8 \\ \hline \end{array}$$

$$\begin{array}{r} \overline{1}\ \ \overline{4} \\ -\quad 8 \\ \hline \end{array}$$

$$\begin{array}{r} \overline{1}\ \ \overline{6} \\ -\quad 7 \\ \hline \end{array}$$

$$\begin{array}{r} \overline{1}\ \ \overline{3} \\ -\quad 9 \\ \hline \end{array}$$

$$\begin{array}{r} \overline{1}\ \ \overline{8} \\ -\quad 9 \\ \hline \end{array}$$

$$\begin{array}{r} \overline{1}\ \ \overline{2} \\ -\quad 8 \\ \hline \end{array}$$

$$\begin{array}{r} \overline{1}\ \ \overline{5} \\ -\quad 9 \\ \hline \end{array}$$

$$\begin{array}{r} \overline{1}\ \ \overline{3} \\ -\quad 9 \\ \hline \end{array}$$

$$\begin{array}{r} \overline{1}\ \ \overline{5} \\ -\quad 8 \\ \hline \end{array}$$

2. <u>Solve</u> the problem and <u>write</u> the missing number:

10 candies   =   5 cupcakes

2 candies = __ cupcake(s)

Answer: _____.

1. <u>Add.</u>   Score ___/15   Time __:__

$$\begin{array}{r} \bar{} \ 4 \\ + \ 9 \\ \hline \end{array}$$ $$\begin{array}{r} \bar{} \ 5 \\ + \ 8 \\ \hline \end{array}$$ $$\begin{array}{r} \bar{} \ 9 \\ + \ 7 \\ \hline \end{array}$$ $$\begin{array}{r} \bar{} \ 5 \\ + \ 5 \\ \hline \end{array}$$ $$\begin{array}{r} \bar{} \ 6 \\ + \ 8 \\ \hline \end{array}$$

$$\begin{array}{r} \bar{} \ 4 \\ + \ 7 \\ \hline \end{array}$$ $$\begin{array}{r} \bar{} \ 7 \\ + \ 7 \\ \hline \end{array}$$ $$\begin{array}{r} \bar{} \ 4 \\ + \ 8 \\ \hline \end{array}$$ $$\begin{array}{r} \bar{} \ 6 \\ + \ 7 \\ \hline \end{array}$$ $$\begin{array}{r} \bar{} \ 9 \\ + \ 3 \\ \hline \end{array}$$

$$\begin{array}{r} \bar{} \ 8 \\ + \ 9 \\ \hline \end{array}$$ $$\begin{array}{r} \bar{} \ 2 \\ + \ 8 \\ \hline \end{array}$$ $$\begin{array}{r} \bar{} \ 5 \\ + \ 9 \\ \hline \end{array}$$ $$\begin{array}{r} \bar{} \ 3 \\ + \ 9 \\ \hline \end{array}$$ $$\begin{array}{r} \bar{} \ 5 \\ + \ 8 \\ \hline \end{array}$$

2.

I have some numbers and signs: 2, 3, 3, +, -.

<u>Write</u> the equation that equals one of the answer choices.

A 1   C 4

B 5   D 7   _____

               www.homerunpress.com

1. <u>Subtract.</u>　　　Score ___/15　　Time __:__

$$
\begin{array}{r} 1\ 6 \\ -\ \ \ 9 \\ \hline \end{array}
\qquad
\begin{array}{r} 1\ 2 \\ -\ \ \ 8 \\ \hline \end{array}
\qquad
\begin{array}{r} 1\ 4 \\ -\ \ \ 7 \\ \hline \end{array}
\qquad
\begin{array}{r} 1\ 1 \\ -\ \ \ 5 \\ \hline \end{array}
\qquad
\begin{array}{r} 1\ 5 \\ -\ \ \ 8 \\ \hline \end{array}
$$

$$
\begin{array}{r} 1\ 1 \\ -\ \ \ 7 \\ \hline \end{array}
\qquad
\begin{array}{r} 1\ 5 \\ -\ \ \ 7 \\ \hline \end{array}
\qquad
\begin{array}{r} 1\ 1 \\ -\ \ \ 8 \\ \hline \end{array}
\qquad
\begin{array}{r} 1\ 3 \\ -\ \ \ 7 \\ \hline \end{array}
\qquad
\begin{array}{r} 1\ 5 \\ -\ \ \ 9 \\ \hline \end{array}
$$

$$
\begin{array}{r} 1\ 2 \\ -\ \ \ 9 \\ \hline \end{array}
\qquad
\begin{array}{r} 1\ 3 \\ -\ \ \ 8 \\ \hline \end{array}
\qquad
\begin{array}{r} 1\ 4 \\ -\ \ \ 9 \\ \hline \end{array}
\qquad
\begin{array}{r} 1\ 7 \\ -\ \ \ 9 \\ \hline \end{array}
\qquad
\begin{array}{r} 1\ 4 \\ -\ \ \ 8 \\ \hline \end{array}
$$

2.

I have some numbers and signs: 4, 5, 7, +, -.

<u>Write</u> the equation that equals one of the answer choices.

A　4　　　C　10

B　6　　　D　9　_____

1. <u>Add.</u> Score ___/15 Time __:__

```
 ‾ 9      ‾ 6      ‾ 7      ‾ 8      ‾ 5
+  9     +  6     +  7     +  8     +  5
_____   _____   _____   _____   _____
```

```
 ‾ 4      ‾ 3      ‾ 4      ‾ 5      ‾ 4
+  9     +  7     +  8     +  7     +  7
_____   _____   _____   _____   _____
```

```
 ‾ 6      ‾ 3      ‾ 8      ‾ 5      ‾ 3
+  9     +  9     +  9     +  9     +  9
_____   _____   _____   _____   _____
```

2.

I have some numbers and signs: 12, 7, -.

<u>Write</u> the equation that equals one of the answer choices.

A 6   C 7

B 4   D 5   _____

www.homerunpress.com

1.

Solve the problem:

9728: the difference of the thousands and the tens is _____.

Answer:

_____

1. <u>Subtract.</u>   Score ___/15   Time __:__

| 1 6 | 1 2 | 1 4 | 1 1 | 1 5 |
| - 7 | - 3 | - 5 | - 4 | - 6 |

| 1 1 | 1 5 | 1 1 | 1 3 | 1 5 |
| - 3 | - 8 | - 7 | - 6 | - 8 |

| 1 2 | 1 3 | 1 4 | 1 7 | 1 4 |
| - 6 | - 5 | - 7 | - 8 | - 9 |

2.

I have some numbers and signs: 17, 8, -. <u>Write</u> the equation that equals one of the answer choices.

A   8   C   9

B   11   D   7   _____

www.homerunpress.com

1. <u>Add.</u>   Score ___/15   Time __:__

```
−  6      −  6      −  6      −  6      −  6
+  9      +  6      +  7      +  8      +  4
_____    _____    _____    _____    _____


−  8      −  8      −  8      −  8      −  8
+  6      +  2      +  5      +  7      +  8
_____    _____    _____    _____    _____


−  7      −  3      −  7      −  7      −  7
+  9      +  9      +  3      +  6      +  5
_____    _____    _____    _____    _____
```

2.

I have some numbers and signs: 9, 4, +.

<u>Write</u> the equation that equals one of the answer choices.

A 12   C 14
B 13   D 11   _____

1. <u>Subtract.</u>   Score ___/15   Time __:__

$$\begin{array}{r} 1\ 1 \\ -\quad 7 \\ \hline \end{array}$$ $$\begin{array}{r} 1\ 1 \\ -\quad 3 \\ \hline \end{array}$$ $$\begin{array}{r} 1\ 1 \\ -\quad 5 \\ \hline \end{array}$$ $$\begin{array}{r} 1\ 1 \\ -\quad 4 \\ \hline \end{array}$$ $$\begin{array}{r} 1\ 1 \\ -\quad 6 \\ \hline \end{array}$$

$$\begin{array}{r} 1\ 2 \\ -\quad 3 \\ \hline \end{array}$$ $$\begin{array}{r} 1\ 2 \\ -\quad 8 \\ \hline \end{array}$$ $$\begin{array}{r} 1\ 2 \\ -\quad 7 \\ \hline \end{array}$$ $$\begin{array}{r} 1\ 2 \\ -\quad 6 \\ \hline \end{array}$$ $$\begin{array}{r} 1\ 2 \\ -\quad 8 \\ \hline \end{array}$$

$$\begin{array}{r} 1\ 3 \\ -\quad 6 \\ \hline \end{array}$$ $$\begin{array}{r} 1\ 3 \\ -\quad 5 \\ \hline \end{array}$$ $$\begin{array}{r} 1\ 3 \\ -\quad 7 \\ \hline \end{array}$$ $$\begin{array}{r} 1\ 3 \\ -\quad 8 \\ \hline \end{array}$$ $$\begin{array}{r} 1\ 3 \\ -\quad 9 \\ \hline \end{array}$$

2.

I have some numbers and signs: 3, 3, 5, +, +.

<u>Write</u> the equation that equals one of the answer choices.

A   10   C   13
B   12   D   11   _____

www.homerunpress.com

1. <u>Subtract.</u>   Score ___/15   Time __:__

$$\begin{array}{r} 16 \\ -\ \ 8 \\ \hline \end{array} \qquad \begin{array}{r} 12 \\ -\ \ 5 \\ \hline \end{array} \qquad \begin{array}{r} 14 \\ -\ \ 9 \\ \hline \end{array} \qquad \begin{array}{r} 11 \\ -\ \ 2 \\ \hline \end{array} \qquad \begin{array}{r} 15 \\ -\ \ 7 \\ \hline \end{array}$$

$$\begin{array}{r} 11 \\ -\ \ 6 \\ \hline \end{array} \qquad \begin{array}{r} 15 \\ -\ \ 9 \\ \hline \end{array} \qquad \begin{array}{r} 11 \\ -\ \ 4 \\ \hline \end{array} \qquad \begin{array}{r} 13 \\ -\ \ 8 \\ \hline \end{array} \qquad \begin{array}{r} 15 \\ -\ \ 8 \\ \hline \end{array}$$

$$\begin{array}{r} 12 \\ -\ \ 4 \\ \hline \end{array} \qquad \begin{array}{r} 13 \\ -\ \ 6 \\ \hline \end{array} \qquad \begin{array}{r} 14 \\ -\ \ 6 \\ \hline \end{array} \qquad \begin{array}{r} 17 \\ -\ \ 9 \\ \hline \end{array} \qquad \begin{array}{r} 14 \\ -\ \ 6 \\ \hline \end{array}$$

2.

Circle the right answer:

I have a series of numbers: 3, 6, 5, 8, 7, __.

What is the next number?

A 10   C 11
B 12   D 9

1. Add.   Score ___/15   Time __:__

$$
\begin{array}{r} 7 \\ + \ 9 \\ \hline \end{array}
\qquad
\begin{array}{r} 6 \\ + \ 8 \\ \hline \end{array}
\qquad
\begin{array}{r} 4 \\ + \ 7 \\ \hline \end{array}
\qquad
\begin{array}{r} 6 \\ + \ 9 \\ \hline \end{array}
\qquad
\begin{array}{r} 8 \\ + \ 4 \\ \hline \end{array}
$$

$$
\begin{array}{r} 7 \\ + \ 6 \\ \hline \end{array}
\qquad
\begin{array}{r} 8 \\ + \ 5 \\ \hline \end{array}
\qquad
\begin{array}{r} 9 \\ + \ 5 \\ \hline \end{array}
\qquad
\begin{array}{r} 8 \\ + \ 9 \\ \hline \end{array}
\qquad
\begin{array}{r} 7 \\ + \ 8 \\ \hline \end{array}
$$

$$
\begin{array}{r} 7 \\ + \ 3 \\ \hline \end{array}
\qquad
\begin{array}{r} 2 \\ + \ 9 \\ \hline \end{array}
\qquad
\begin{array}{r} 7 \\ + \ 5 \\ \hline \end{array}
\qquad
\begin{array}{r} 4 \\ + \ 6 \\ \hline \end{array}
\qquad
\begin{array}{r} 7 \\ + \ 7 \\ \hline \end{array}
$$

2.

Circle the right answer:

I have a series of numbers: 5, 11, 7, 13, 9, __.

A 19   C 15

B 13   D 17

1. <u>Add.</u>   Score ___/15   Time __:__

$$\begin{array}{r} ^-\ 9 \\ +\ 9 \\ \hline \end{array} \qquad \begin{array}{r} ^-\ 4 \\ +\ 8 \\ \hline \end{array} \qquad \begin{array}{r} ^-\ 7 \\ +\ 7 \\ \hline \end{array} \qquad \begin{array}{r} ^-\ 6 \\ +\ 5 \\ \hline \end{array} \qquad \begin{array}{r} ^-\ 8 \\ +\ 8 \\ \hline \end{array}$$

$$\begin{array}{r} ^-\ 9 \\ +\ 6 \\ \hline \end{array} \qquad \begin{array}{r} ^-\ 5 \\ +\ 5 \\ \hline \end{array} \qquad \begin{array}{r} ^-\ 9 \\ +\ 7 \\ \hline \end{array} \qquad \begin{array}{r} ^-\ 8 \\ +\ 7 \\ \hline \end{array} \qquad \begin{array}{r} ^-\ 9 \\ +\ 8 \\ \hline \end{array}$$

$$\begin{array}{r} ^-\ 7 \\ +\ 5 \\ \hline \end{array} \qquad \begin{array}{r} ^-\ 2 \\ +\ 8 \\ \hline \end{array} \qquad \begin{array}{r} ^-\ 9 \\ +\ 5 \\ \hline \end{array} \qquad \begin{array}{r} ^-\ 8 \\ +\ 6 \\ \hline \end{array} \qquad \begin{array}{r} ^-\ 7 \\ +\ 8 \\ \hline \end{array}$$

2.

I have some numbers and signs: 6, 7, 11, +, -.

<u>Write</u> the equation that equals one of the answer choices.

A   13   C   4
B   2   D   16   _____

1. <u>Subtract.</u>   Score ___/15   Time __:__

$$\begin{array}{r} 1\ 6 \\ -\ \ 8 \\ \hline \end{array}$$ $$\begin{array}{r} 1\ 2 \\ -\ \ 6 \\ \hline \end{array}$$ $$\begin{array}{r} 1\ 4 \\ -\ \ 7 \\ \hline \end{array}$$ $$\begin{array}{r} 1\ 1 \\ -\ \ 5 \\ \hline \end{array}$$ $$\begin{array}{r} 1\ 5 \\ -\ \ 7 \\ \hline \end{array}$$

$$\begin{array}{r} 1\ 1 \\ -\ \ 4 \\ \hline \end{array}$$ $$\begin{array}{r} 1\ 7 \\ -\ \ 8 \\ \hline \end{array}$$ $$\begin{array}{r} 1\ 2 \\ -\ \ 7 \\ \hline \end{array}$$ $$\begin{array}{r} 1\ 3 \\ -\ \ 8 \\ \hline \end{array}$$ $$\begin{array}{r} 1\ 7 \\ -\ \ 8 \\ \hline \end{array}$$

$$\begin{array}{r} 1\ 2 \\ -\ \ 3 \\ \hline \end{array}$$ $$\begin{array}{r} 1\ 3 \\ -\ \ 9 \\ \hline \end{array}$$ $$\begin{array}{r} 1\ 4 \\ -\ \ 8 \\ \hline \end{array}$$ $$\begin{array}{r} 1\ 7 \\ -\ \ 9 \\ \hline \end{array}$$ $$\begin{array}{r} 1\ 4 \\ -\ \ 5 \\ \hline \end{array}$$

2.

I have some numbers and signs: 4, 6, 8, +, -.

<u>Write</u> the equation that equals one of the answer choices.

A 9   C 6
B 7   D 12   _____

www.homerunpress.com

Page 5

1. Read

When I add 3 candies and 2 candies, there are 5 candies altogether. It does not matter which way I add candies together.

3 + 2 = 5 candies

means equals   means add or plus

2 + 3 = 5 candies

I add 4 cars and 3 trucks.

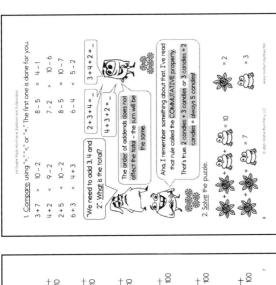

I have 4 cars and 3 trucks together. I can find the total simply by counting them all. There are 7 in all.

4 + 3 = 7

Page 6

1. Read

I use a number line to find out the answer when I add 4 and 2. First, I draw a line and mark it with numbers. I find 4 on the number line.

Start counting at 4.

I need to add 2, so I jump 2 places to the right.

This takes me to 6.

So   4 + 2 = 6

I add 30 and 50. First, I find 30. Then, I jump 5 places to the right.

So   30 + 50 = 80

Page 7

1. Add. Use a number line to show the jumps.

2 + 7 = 9

6 + 4 = 10

9 + 1 = 10

50 + 50 = 100

40 + 40 = 80

30 + 70 = 100

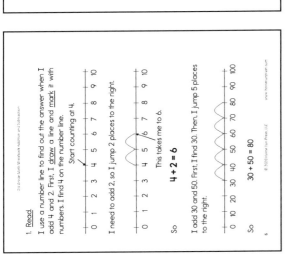

Page 8

1. Compare using ">", "<", or "=". The first one is done for you.

3 + 7 > 10 - 2   8 - 5 > 4 - 1
4 + 2 > 9 - 2   7 - 2 > 10 - 6
2 + 5 < 10 - 2   8 - 5 = 10 - 7
6 + 3 > 4 + 3   6 - 4 < 5 - 2

"We need to add 3, 4 and 2. What is the total?"

2 + 3 + 4 = ___   →   3 + 4 + 2 = ___
4 + 3 + 2 = ___

"The order of addends does not affect the total – the sum will be the same."

"Aha, I remember something about that. I've read that rule called the COMMUTATIVE property. That's true. 2 candles + 3 candles or 3 candles + 2 candles = always 5 candles!"

2. Solve the puzzle.

flower + frog = 10

frog + frog = ___

flower + flower = ___

frog = 2

flower = 3

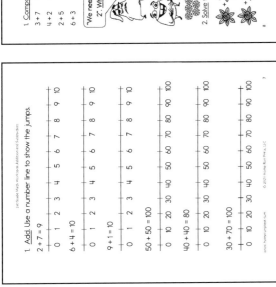

Page 9

1. Add and change the addends' order. The first one is done for you. Answers will vary.

2 + 5 + 3 = 10   5 + 3 + 2 = 10   3 + 2 + 5 = 10
1 + 4 + 2 = 7   _ + _ + _ = _   _ + _ + _ = _
5 + 0 + 5 = 10   _ + _ + _ = _   _ + _ + _ = _
4 + 1 + 3 = 8   _ + _ + _ = _   _ + _ + _ = _
6 + 1 + 3 = 10   _ + _ + _ = _   _ + _ + _ = _

2. Complete an addition number sentence with tens and ones.

16 = 10 + 6   18 = 10 + 8   15 = 10 + 5
10 = 10 + 0   13 = 10 + 3   17 = 10 + 7

3. Write the missing numbers to make the comparison true.

1 + 9 = 14 - 4   9 - 4 = 7 - 2
3 + 4 = 10 - 3   10 - 6 = 1 + 3
4 + 4 = 6 + 2   6 - 5 = 8 - 7

Page 10

"I'm faster than Pickles, but Sunny is faster than me. Who is the fastest?"

Answer: Sunny > I > Pickles

Page 11

1. The blocks in each tower tells you how many hundreds, tens, and ones in each number. Write and put the numbers in order from the least to the greatest.

214   430   502

2. Write and put the numbers in order from the largest to the smallest.

313   142   124

3. My sister found 9 apples. She ate 6 apples. How many apples are left?

Answer: 9 - 6 = 3

Page 12

1. Make and write the smallest and the biggest two-digit numbers you can with any two of these digits: 7, 4, 6.

Answer: 9467, 7641

2. Make and write the smallest and the biggest two-digit numbers you can with any two of these digits: 9, 5, 9, 8.

Answer: 5899, 9985

3. Make and write the smallest and the biggest two-digit numbers you can with any two of these digits: 2, 9, 3, 0.

Answer: 2039, 9320

4. Write the missing number.

500 + 40 + 8 = 548   100 + 20 + 6 = 126
200 + 50 + 1 = 251   300 + 80 + 9 = 389
600 + 10 + 0 = 610   400 + 80 + 4 = 484
700 + 30 + 7 = 737   800 + 60 + 2 = 862
900 + 10 + 1 = 911   500 + 40 + 1 = 541

Page 13

1. Read

I like to split the adding numbers into numbers that are easier to work with. I can show my favorite strategy. T = tens, O = ones.

Step 1. Let's add 12 and 15.

T O	T O
12 + 15 = ___	

Step 2. Add the tens together.

T O	T O
10 + 10 = 20	

Step 3. Add the ones together.

T O	T O
2 + 5 = 7	

Step 4. Add the tens and ones to find the total.

T	O	T O
20 + 7 = 27		

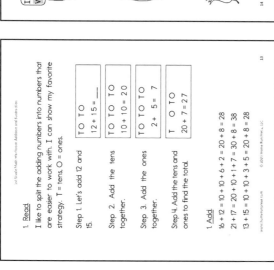

1. Add

16 + 12 = 10 + 10 + 6 + 2 = 20 + 8 = 28

21 + 17 = 20 + 10 + 1 + 7 = 30 + 8 = 38

13 + 15 = 10 + 10 + 3 + 5 = 20 + 8 = 28

www.homerunpress.com   © 2021 Home Run Press, LLC

Page 14

1st Grade Math Workbook Addition and Subtraction

I can add numbers using column addition. Hint: Write ones under ones. Write tens under tens.

Step 1. Write the digits that have the same place value lined up one above the other.

tens	ones
1	2
+ 1	5
	—

Step 2. Start by adding the ones together. Add 2 ones and 5 ones: 2 + 5 = 7. Write 7 in the ones column.

tens	ones
1	2
+ 1	5
	7

Step 3. Add 1 ten and 1 ten. But I actually add 10 and 10. So the answer is: 10 + 10 = 20. I write 2 in the tens column.

tens	ones
1	2
+ 1	5
2	7

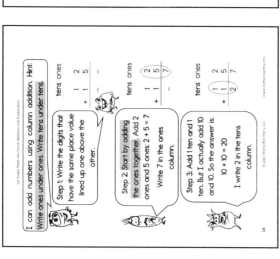

© 2021 Home Run Press, LLC   www.homerunpress.com

Page 15

1st Grade Math Workbook Addition and Subtraction

1. Add.

$$\begin{array}{r}3\\+3\\\hline 6\end{array}\quad\begin{array}{r}5\\+2\\\hline 7\end{array}\quad\begin{array}{r}3\\+6\\\hline 9\end{array}\quad\begin{array}{r}7\\+2\\\hline 9\end{array}\quad\begin{array}{r}4\\+4\\\hline 8\end{array}\quad\begin{array}{r}8\\+1\\\hline 9\end{array}$$

$$\begin{array}{r}2\\+8\\\hline 10\end{array}\quad\begin{array}{r}5\\+5\\\hline 10\end{array}\quad\begin{array}{r}3\\+7\\\hline 10\end{array}\quad\begin{array}{r}4\\+6\\\hline 10\end{array}\quad\begin{array}{r}1\\+9\\\hline 10\end{array}\quad\begin{array}{r}6\\+4\\\hline 10\end{array}$$

2. Complete each pair of number bonds.

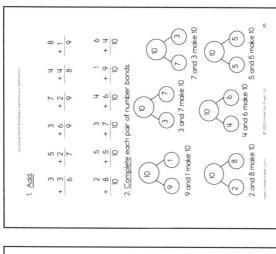

9 and 1 make 10   3 and 7 make 10

2 and 8 make 10   4 and 6 make 10

7 and 3 make 10   5 and 5 make 10

© 2022 Home Run Press, LLC   15

Page 16

1st Grade Math Workbook Addition and Subtraction

1. Use cupcakes to make 10. Color the cupcakes brown and yellow.

1 + 9 = 10

9 + 1 = 10

1 + 9 and 9 + 1 are two ways of making 10

2 + 8 = 10

8 + 2 = 10

3 + 7 = 10

7 + 3 = 10

4 + 6 = 10

6 + 4 = 10

© 2022 Home Run Press, LLC   16

Page 17

1st Grade Math Workbook Addition and Subtraction

1. Complete each picture. When a shape is symmetrical, each half is a mirror image of the other.

Symmetry

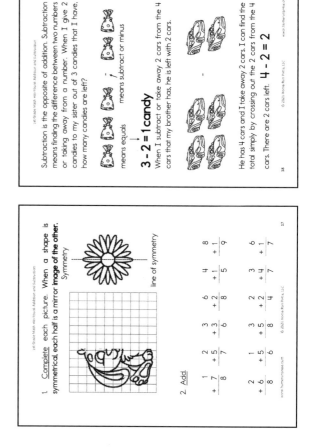

line of symmetry

2. Add

$$\begin{array}{r}1\\+7\\\hline 8\end{array}\quad\begin{array}{r}2\\+5\\\hline 7\end{array}\quad\begin{array}{r}3\\+3\\\hline 6\end{array}\quad\begin{array}{r}6\\+2\\\hline 8\end{array}\quad\begin{array}{r}4\\+1\\\hline 5\end{array}\quad\begin{array}{r}8\\+1\\\hline 9\end{array}$$

$$\begin{array}{r}2\\+6\\\hline 8\end{array}\quad\begin{array}{r}1\\+5\\\hline 6\end{array}\quad\begin{array}{r}3\\+5\\\hline 8\end{array}\quad\begin{array}{r}2\\+2\\\hline 4\end{array}\quad\begin{array}{r}3\\+4\\\hline 7\end{array}\quad\begin{array}{r}6\\+1\\\hline 7\end{array}$$

© 2021 Home Run Press, LLC   17

Page 18

1st Grade Math Workbook Addition and Subtraction

Subtraction is the opposite of addition. Subtraction means finding the difference between two numbers or taking away from a number. When I give 2 candies to my sister out of 3 candies that I have, how many candies are left?

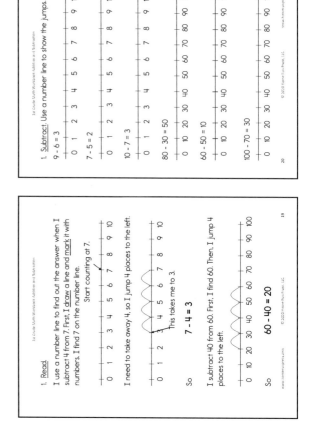

means equals   means subtract or minus

3 − 2 = 1 candy

When I subtract or take away 2 cars from the 4 cars that my brother has, he is left with 2 cars.

He has 4 cars and I take away 2 cars. I can find the total simply by crossing out the 2 cars from the 4 cars. There are 2 cars left. **4 − 2 = 2**

www.homerunpress.com   © 2021 Home Run Press, LLC

Page 19

1st Grade Math Workbook Addition and Subtraction

1. Read.

I use a number line to find out the answer when I subtract 4 from 7. First, I draw a line and mark it with numbers. I find 7 on the number line.

Start counting at 7.

0 1 2 3 4 5 6 7 8 9 10

I need to take away 4, so I jump 4 places to the left.

0 1 2 3 4 5 6 7 8 9 10

This takes me to 3.

So   7 − 4 = 3

I subtract 40 from 60. First, I find 60. Then, I jump 4 places to the left.

0 10 20 30 40 50 60 70 80 90 100

So   60 − 40 = 20

www.homerunpress.com   © 2021 Home Run Press, LLC   19

Page 20

1st Grade Math Workbook Addition and Subtraction

1. Subtract: Use a number line to show the jumps.

9 − 6 = 3

0 1 2 3 4 5 6 7 8 9 10

7 − 5 = 2

0 1 2 3 4 5 6 7 8 9 10

10 − 7 = 3

0 1 2 3 4 5 6 7 8 9 10

80 − 30 = 50

0 10 20 30 40 50 60 70 80 90 100

60 − 50 = 10

0 10 20 30 40 50 60 70 80 90 100

100 − 70 = 30

0 10 20 30 40 50 60 70 80 90 100

© 2022 Home Run Press, LLC   20

Page 21

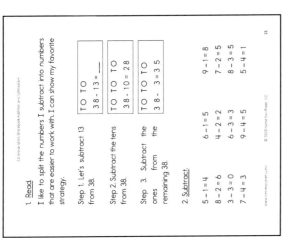

1. Read.
I like to split the numbers I subtract into numbers that are easier to work with. I can show my favorite strategy.

Step 1. Let's subtract 13 from 38.

Step 2. Subtract the tens from 38.

Step 3. Subtract the ones from the remaining 38.

| T O | T O |
| $38 - 13 =$ ___ |

| T O | T O |
| $38 - 10 = 28$ |

| T O | T O |
| $38 - 3 = 35$ |

2. Subtract.

$5 - 1 = 4$ $6 - 1 = 5$ $9 - 1 = 8$
$8 - 2 = 6$ $4 - 2 = 2$ $7 - 2 = 5$
$3 - 3 = 0$ $6 - 3 = 3$ $8 - 3 = 5$
$7 - 4 = 3$ $9 - 4 = 5$ $5 - 4 = 1$

Page 22

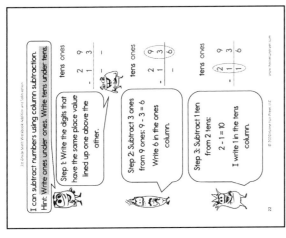

I can subtract numbers using column subtraction.
Hint: Write ones under ones. Write tens under tens.

Step 1: Write the digits that have the same place value lined up one above the other.

Step 2: Subtract 3 ones from 9 ones: $9 - 3 = 6$
Write 6 in the ones column.

Step 3: Subtract 1 ten from 2 tens:
$2 - 1 = 10$
I write 1 in the tens column.

Page 23

1. Subtract.

$9 - 3 = 6$ $5 - 2 = 3$ $8 - 4 = 4$ $7 - 2 = 5$ $4 - 4 = 0$ $8 - 1 = 7$

$8 - 2 = 6$ $5 - 5 = 0$ $5 - 2 = 3$ $7 - 6 = 1$ $9 - 3 = 6$ $6 - 4 = 2$

2. Complete each pair of number bonds.

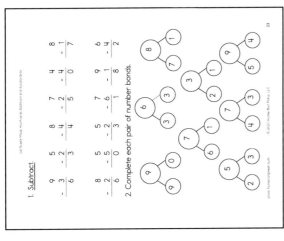

Page 24

1. Subtract.

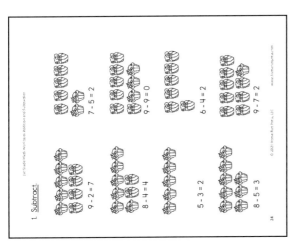

$9 - 2 = 7$ $7 - 5 = 2$

$8 - 4 = 4$ $9 - 9 = 0$

$5 - 3 = 2$ $6 - 4 = 2$

$8 - 5 = 3$ $9 - 7 = 2$

Page 25

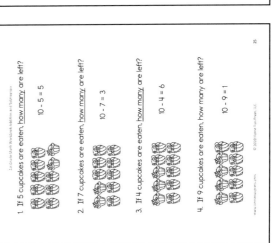

1. If 5 cupcakes are eaten, how many are left?
$10 - 5 = 5$

2. If 7 cupcakes are eaten, how many are left?
$10 - 7 = 3$

3. If 4 cupcakes are eaten, how many are left?
$10 - 4 = 6$

4. If 9 cupcakes are eaten, how many are left?
$10 - 9 = 1$

Page 26

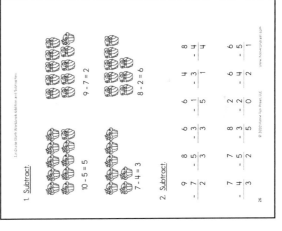

1. Subtract.

$10 - 5 = 5$ $9 - 7 = 2$

$7 - 4 = 3$ $8 - 2 = 6$

2. Subtract.

$9 - 2 = 7$ $8 - 3 = 5$ $6 - 1 = 5$ $4 - 3 = 1$ $8 - 4 = 4$

$7 - 4 = 3$ $7 - 5 = 2$ $8 - 3 = 5$ $6 - 2 = 4$ $6 - 5 = 1$

Page 27

1. Color the flowers with the smaller number in each group.

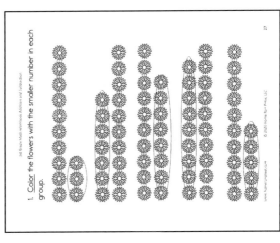

Page 28

1. Subtract.

$10 - 3 = 7$ $10 - 5 = 5$ $10 - 6 = 4$ $10 - 8 = 2$ $10 - 4 = 6$ $10 - 1 = 9$

$17 - 6 = 11$ $18 - 5 = 13$ $15 - 1 = 14$ $16 - 2 = 14$ $19 - 1 = 18$

2. Complete each pair of number bonds.

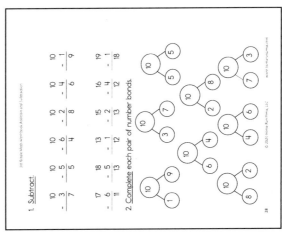

Page 29

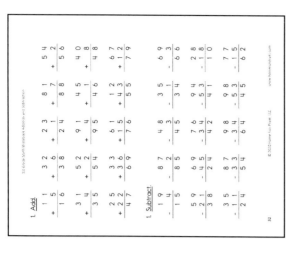

1. Read. Write how many tens. Write the number.
Two digit numbers are made up of tens and ones.

1 ten, 10

3 tens, 30

5 tens, 50

4 tens, 40

8 tens, 80

Page 30

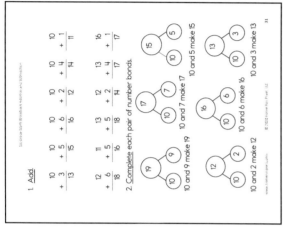

1. How many are there in each group?

10 1
10 + 1 makes 11

10 2
10 + 2 makes 12

10 3
10 + 3 makes 13

10 4
10 + 4 makes 14

10 5
10 + 5 makes 15

Page 31

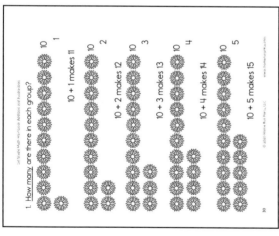

1. Add.

10 +3 = 13	10 +5 = 15	10 +6 = 16	10 +2 = 12	10 +4 = 14	10 +1 = 11
12 +6 = 18	11 +5 = 16	13 +5 = 18	12 +2 = 14	13 +4 = 17	16 +1 = 17

2. Complete each pair of number bonds.

- 19 → 10, 9 — 10 and 9 make 19
- 17 → 10, 7 — 10 and 7 make 17
- 15 → 10, 5 — 10 and 5 make 15
- 12 → 10, 2 — 10 and 2 make 12
- 16 → 10, 6 — 10 and 6 make 16
- 13 → 10, 3 — 10 and 3 make 13

Page 32

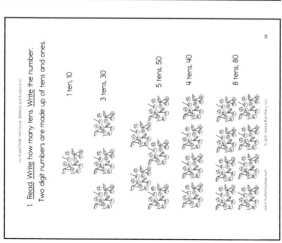

1. Add.

(columns of two-digit addition problems)

1. Subtract.

(columns of two-digit subtraction problems)

Page 33

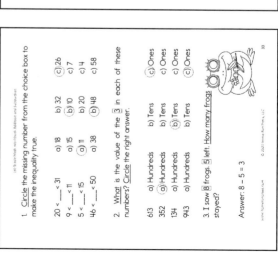

1. Circle the missing number from the choice box to make the inequality true.

- 20 < ___ < 31   a) 18   b) 32   c) 26
- 9 < ___ < 11   a) 15   b) 10   c) 7
- 5 < ___ < 15   a) 11   b) 20   c) 4
- 46 < ___ < 50   a) 38   b) 48   c) 58

2. What is the value of the 3 in each of these numbers? Circle the right answer.

- 613   a) Hundreds   b) Tens   c) Ones
- 352   a) Hundreds   b) Tens   c) Ones
- 134   a) Hundreds   b) Tens   c) Ones
- 943   a) Hundreds   b) Tens   c) Ones

3. I saw 8 frogs. 5 left. How many frogs stayed?

Answer: 8 − 5 = 3

Page 34

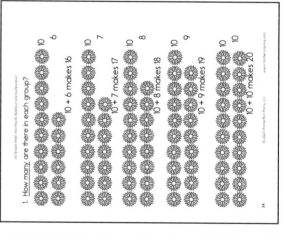

1. How many are there in each group?

10 6
10 + 6 makes 16

10 7
10 + 7 makes 17

10 8
10 + 8 makes 18

10 9
10 + 9 makes 19

10 10
10 + 10 makes 20

Page 35

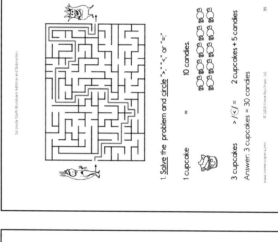

1. Solve the problem and circle ">", "<", or "=":

1 cupcake = 10 candies.

3 cupcakes   >/<=   2 cupcakes + 5 candies

Answer: 3 cupcakes = 30 candies

Page 36

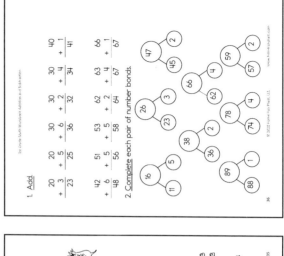

1. Add.

20 +3 = 23	20 +5 = 25	30 +6 = 36	30 +2 = 32	30 +4 = 34	40 +1 = 41
42 +6 = 48	51 +5 = 56	53 +5 = 58	62 +2 = 64	63 +4 = 67	66 +1 = 67

2. Complete each pair of number bonds.

- 47 → 45, 2
- 66 → 62, 4   59 → 57, 2
- 26 → 23, 3   78 → 74, 4
- 38 → 36, 2
- 16 → 11, 5   89 → 88, 1

Page 37 (bottom left)

1. The blocks in each tower tell you how many hundreds, tens, and ones in each number. Write and put the numbers in order from the least to the

H	T	O
334   432   513

2. Write the number for these.

2 tens + 5 ones = 25   1 ten + 9 ones = 19
3 tens + 0 ones = 30   5 tens + 6 ones = 56
4 tens + 9 ones = 49   6 tens + 2 ones = 62
7 tens + 0 ones = 70   8 tens + 5 ones = 85

3. What is the value of the digit 7 in the numbers below?

37   71   7   79

ones   tens   ones   ones   tens

Page 38 (left)

1. Add.

$11 + 12 = 23$ $14 + 14 = 28$ $15 + 21 = 36$ $17 + 31 = 48$ $13 + 14 = 27$ $12 + 16 = 28$

$22 + 12 = 34$ $41 + 41 = 82$ $33 + 25 = 58$ $82 + 14 = 96$ $73 + 26 = 99$ $56 + 13 = 69$

2. Complete each pair of number bonds.

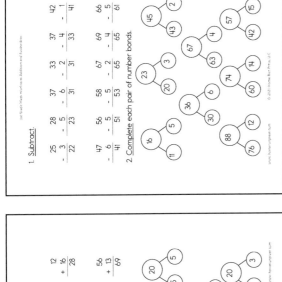

Page 39 (left)

1. Subtract.

$25 - 3 = 22$ $28 - 5 = 23$ $37 - 6 = 31$ $33 - 2 = 31$ $37 - 4 = 33$ $42 - 1 = 41$

$47 - 6 = 41$ $56 - 5 = 51$ $58 - 5 = 53$ $67 - 2 = 65$ $69 - 4 = 65$ $66 - 5 = 61$

2. Complete each pair of number bonds.

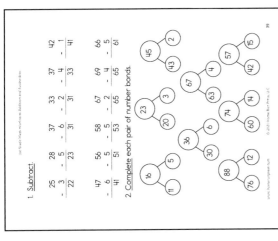

Page 40 (top left)

1. Subtract.

$34 - 12 = 22$ $28 - 14 = 14$ $45 - 21 = 24$ $37 - 31 = 6$ $57 - 14 = 43$ $47 - 16 = 31$

$22 - 12 = 10$ $64 - 41 = 23$ $57 - 25 = 32$ $39 - 14 = 25$ $48 - 26 = 22$ $33 - 11 = 22$

2. Complete each pair of number bonds.

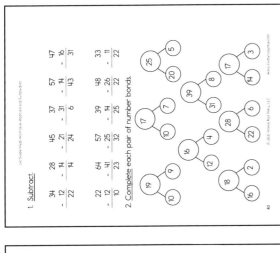

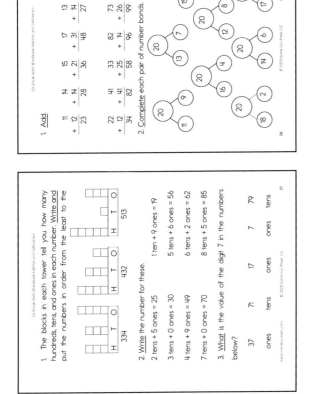

Page 41 (bottom right)

1. How many more do I need to add to the second group to make each group the same?

+6   +7   +5

2. Circle all the combinations that equal 5.

10 - 5   16 - 11   8 - 2

14 - 12   21 - 10   29 - 24

38 - 28   17 - 12   46 - 40

Page 42 (right)

1. If 20 pencils are broken, how many are left?
$100 - 20 = 80$

2. If 30 pencils are broken, how many are left?
$90 - 30 = 60$

3. If 50 pencils are broken, how many are left?
$80 - 50 = 30$

4. If 40 pencils are broken, how many are left?
$70 - 40 = 30$

Page 43 (right)

1. Venn Diagram: helps you sort things according to their different features.

I have many cards. 3 of them are dragon-type cards. 5 of them are flying-type cards. 2 of them are both dragon- and flying-type cards. 2 of them are neither dragon- nor flying-type cards. How many cards are there? Fill in the diagram.

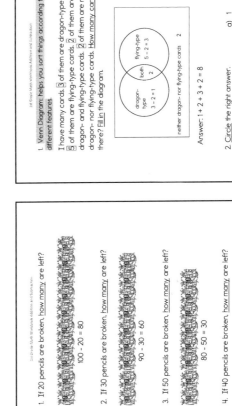

dragon-type $3 - 2 = 1$ both 2   flying-type $5 - 2 = 3$

neither dragon- nor flying-type cards   2

Answer: $1 + 2 + 3 + 2 = 8$

2. Circle the right answer.

a) 1
b) 2
c) 3

_ 2 + 3 6 = 58

Page 44 (top right)

1. Solve the problem and write the missing number:

1 sunflower   =   3 tulips

3 sunflowers =   9 tulips

Answer: $3 + 3 + 3 = 9$.

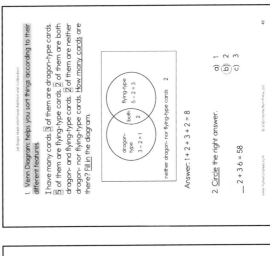
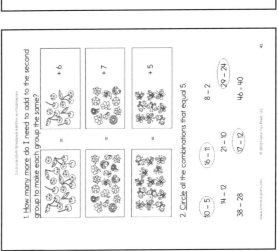

Panel 45

1. **Read**. Write how many tens. **Write** the number.

Three digit numbers are made up of hundreds, tens, and ones.

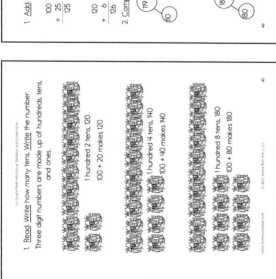

1 hundred 2 tens. 120
100 + 20 makes 120

1 hundred 4 tens. 140
100 + 40 makes 140

1 hundred 8 tens. 180
100 + 80 makes 180

Panel 46

1. **Add**.

100	100	100	100	100
+ 25	+ 15	+ 46	+ 12	+ 47
125	115	146	112	147

120	130	160
+ 47	+ 4	+ 1
147	134	161

120	110	130
+ 6	+ 5	+ 5
126	115	135

2. **Complete** each pair of number bonds.

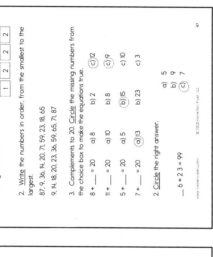

Panel 47

1. I am building a solid slab of rocks: the two rocks next to each other are added to get the number up above. <u>Fill in</u> the missing numbers.

```
        [ 15 ]
     [ 7 ][ 8 ]
   [ 3 ][ 4 ][ 4 ]
 [ 1 ][ 2 ][ 2 ][ 2 ]
```

2. <u>Write</u> the numbers in order, from the smallest to the largest.

87, 9, 36, 14, 20, 71, 59, 23, 18, 65

9, 14, 18, 20, 23, 36, 59, 65, 71, 87

3. Complements to 20. <u>Circle</u> the missing numbers from the choice box to make the equations true.

$8 + \_\_ = 20$ a) 8   b) 2   c) 12

$11 + \_\_ = 20$ a) 10   b) 8   c) 9 (circled)

$5 + \_\_ = 20$ a) 5   b) 15 (circled)   c) 10

$7 + \_\_ = 20$ a) 13 (circled)   b) 23   c) 3

2. <u>Circle</u> the right answer.

$\_\_ 6 + 2\,3 = 99$ a) 5   b) 9   c) 7 (circled)

Panel 48

1. **Read**. Place value is the amount a digit is worth in a number.

Hundreds	Tens	Ones
2	6	3

2 hundreds = the value of 2 in this number is 200
   2 0 0

6 tens = the value of 6 in this number is 60
   6 0

3 ones = the value of 3 in this number is 3
   3

```
  4 0 0
    6 0
+     3   5
```

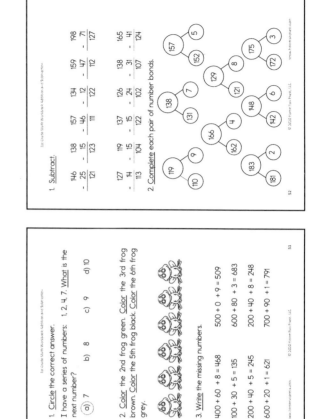

Panel 49

1. <u>What</u> is the value of the 4 in each of these numbers? <u>Circle</u> the right answer.

446   a) Hundreds   b) Tens   c) Ones

914   a) Hundreds   b) Tens   c) Ones (circled)

446   a) Hundreds (circled)   b) Tens   c) Ones

245   a) Hundreds   b) Tens (circled)   c) Ones

624   a) Hundreds   b) Tens   c) Ones (circled)

2. <u>Complete</u> each addition number sentence with tens and ones. The first one is done for you.

45 = 40 + 5	61 = 60 + 1	19 = 10 + 9
30 = 30 + 0	74 = 70 + 4	37 = 30 + 7
52 = 50 + 2	65 = 60 + 5	58 = 50 + 8
98 = 90 + 8	87 = 80 + 7	60 = 60 + 0

Panel 50

1. <u>Write</u> the missing numbers.

50 + 3 = 53	800 + 0 + 7 = 807
20 + 5 = 25	300 + 20 + 8 = 328
90 + 6 = 96	900 + 70 + 2 = 972
60 + 2 = 62	400 + 30 + 0 = 430
10 + 9 = 19	200 + 40 + 4 = 244
40 + 8 = 48	500 + 90 + 5 = 595

2. <u>Circle</u> the correct answer.

I have a series of numbers: 0, 2, 4, 6, 8. <u>What is the</u> next number?

a) 10   b) 8 (circled)   c) 12   d) 7

Panel 51

1. <u>Circle</u> the correct answer.

I have a series of numbers: 1, 2, 4, 7. <u>What is the</u> next number?

a) 7 (circled)   b) 8   c) 9   d) 10

2. <u>Color</u> the 2nd frog green. <u>Color</u> the 3rd frog brown. <u>Color</u> the 5th frog black. <u>Color</u> the 6th frog grey.

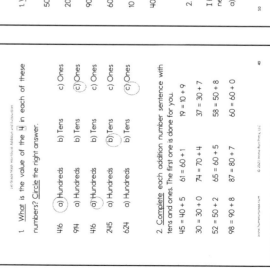

3. <u>Write</u> the missing numbers.

400 + 60 + 8 = 468	500 + 0 + 9 = 509
100 + 30 + 5 = 135	600 + 80 + 3 = 683
200 + 40 + 5 = 245	200 + 40 + 8 = 248
600 + 20 + 1 = 621	700 + 90 + 1 = 791

Panel 52

1. <u>Subtract</u>.

146	138	157	134	159	198
- 25	- 15	- 46	- 12	- 47	- 71
121	123	111	122	112	127

127	119	137	126	138	165
- 14	- 15	- 15	- 24	- 31	- 41
113	104	122	102	107	124

2. <u>Complete</u> each pair of number bonds.

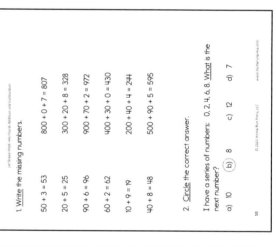

Page 53

1. Complements to 100. Circle the missing numbers from the choice box to make the equations true.

70 + ___ = 100   a) 35   b) 30   c) 10
50 + ___ = 100   a) 40   b) 100   c) 50
90 + ___ = 100   a) 10   b) 20   c) 1

2. Subtract:

```
  3 5       5 4       6 8       8 7
- 0 5     - 0 4     - 0 0     - 0 7
  3 0       5 0       6 0       8 0

  3 5       7 7       8 2       4 9
- 0 4     - 0 2     - 0 1     - 0 4
  3 1       7 5       8 1       4 4

  6 5       9 3       3 9       8 1
- 2 0     - 1 0     - 2 0     - 7 0
  2 5       7 3       1 9       1 1
```

3. The sum of the two 2-digit numbers is 50. Their difference is 30. What are these 2-digit numbers?

Answer: 40 and 10.

40 + 10 = 50   40 - 10 = 30

Page 54

1. Read.

I often need to know if a number is the same as, smaller than, or larger than another number. My teacher calls this comparing numbers. Look at these candies. There are six candies in each row. My teacher says that the number in one row is equal to the number in the second row. 6 = 6

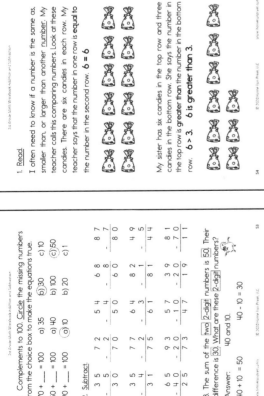

My sister has six candies in the top row and three candies in the bottom row. She says the number in the top row is greater than the number in the bottom row. 6 > 3.   6 is greater than 3.

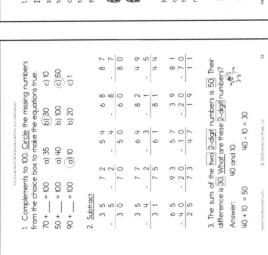

Page 55

1. Read.

My brother has five candies in the top row and six candies in the bottom row. He says the number in the top row is less than the number in the second row. 5 < 6.   5 is less than 6.

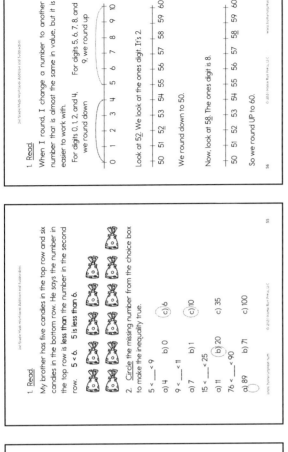

2. Circle the missing number from the choice box to make the inequality true.

5 < ___ < 9
a) 4   b) 0   c) 6

9 < ___ < 11
a) 7   b) 1   c) 10

15 < ___ < 25
a) 11   b) 20   c) 35

76 < ___ < 90
a) 89   b) 71   c) 100

Page 56

1. Read.

When I round, I change a number to another number that is almost the same in value, but it is easier to work with.

For digits 0, 1, 2, and 4, we round down

For digits 5, 6, 7, 8, and 9, we round up

```
0  1  2  3  4  5  6  7  8  9  10
```

Look at 52. We look at the ones digit. It's 2.

We round down to 50.

```
50  51  52  53  54  55  56  57  58  59  60
```

Now, look at 58. The ones digit is 8.

```
50  51  52  53  54  55  56  57  58  59  60
```

So we round UP to 60.

Page 57

1. Which is more? Compare the numbers using ">", "<", or "=".

3 ones < 2 tens   10 ones = 1 ten
3 tens > 8 ones   2 tens > 1 ten
4 tens 2 ones > 4 ones
1 ten 2 ones < 1 ten 20 ones
2 tens 2 ones < 3 tens and 2 ones

2. Round each number to the nearest 10. Look at the next digit to the right. If it is 0,1,2, 3, or 4, then ROUND DOWN, if it is 5, 6, 7, 8, 9, then ROUND UP.

6 10   8 10   5 10
13 10   19 20   16 20
11 10   17 20   19 20
22 20   25 30   23 20
28 30   24 30   29 30

Page 58

1. Which is more? Write the missing numbers to make the comparison true. Answers may vary.

12 ones > 8 ones   7 ones < 1 ten
1 ten = 10 ones   2 tens = 20 ones
5 ones > 3 ones   12 ones < 2 tens
15 ones < 1 ten 6 ones
1 ten 8 ones > 1 ten 5 ones
3 tens and 3 ones > 3 tens and 2 ones
3 tens and 3 ones = 2 tens and 13 ones

2. Round each number to the nearest 100. Look at the next digit to the right. If it is 0, 1, 2, 3, or 4, then ROUND DOWN, if it is 5, 6, 7, 8, 9, then ROUND UP.

153 200   208 200   715 700
913 900   259 300   246 200
371 400   557 600   469 500
622 600   485 500   923 900

Page 59

1. Read.

Even numbers are made of pairs.

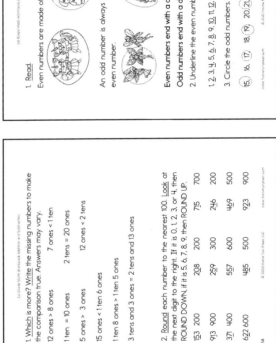

An odd number is always 1 more or 1 less than an even number.

Even numbers end with a digit of 0, 2, 4, 6, 8.
Odd numbers end with a digit of 1, 3, 5, 7, 9.

2. Underline the even numbers.

1, 2, 3, 4, 5, 6, 7, 8, 9, 10, 11, 12, 13, 14, 15, 16

3. Circle the odd numbers.

15, 16, 17, 18, 19, 20, 21, 22, 23, 24, 25, 26

Page 60

1. Read.

I have tons of candies. I need to estimate because it would take too long to count the exact number. I count 5 candies in the bottom row. There are 4 rows, so I can say there are about 5 + 5 + 5 + 5 candies, which is 20 candies.

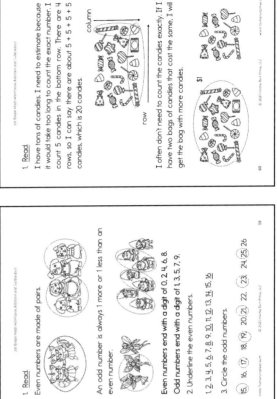

I often don't need to count the candies exactly. If I have two bags of candies that cost the same, I will get the bag with more candies.

Page 64

1. Write the missing numbers to make the scales balance.

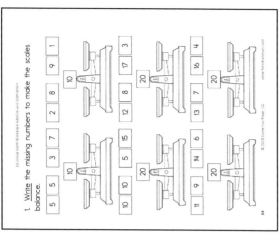

Page 68

1. What number am I?

Half of me is 1 and double me is 4.

One whole.

A half of two equal parts of one whole is one.

Double means take twice as much or as many.

Half of me is 2 and double me is 8.

I am 4.

Half of me is 5 and double me is 20.

I am 10.

Half of me is 10 and double me is 40.

I am 20.

Half of me is 50 and double me is 200.

I am 100.

Page 63

1. I need money to buy objects. Write the missing numbers.

2 coins: 15¢ $10 + 5$

3 coins: 15¢ $5 + 5 + 5$

4 coins: 17¢ $10 + 5 + 1 + 1$

2 coins: 20¢ $10 + 10$

2 coins: 30¢ $25 + 5$

5 coins: 17¢ $5 + 5 + 5 + 1 + 1$

Page 67

1. I got 8 candies. I ate a half of the candies. Color them red. How many candies are left?

Circle your answer: 0 1 2 3 (4) 5

I found 10 flowers. A half of the flowers were blooming. How many flowers were not blooming?

Circle your answer: 0 1 2 3 4 (5)

The pumpkin weighed 2 pounds. We ate a half of it. How many pounds are left?

Circle your answer: 0 (1) 2 3 4 5

My birthday cake weighed 10 pounds! My friends ate a half of the cake. How many pounds are left?

Circle your answer: 0 1 2 3 4 (5)

Page 62

1. I need money to buy objects. Write the missing numbers.

2 coins: 11¢ $10 + 1$

3 coins: 16¢ $10 + 5 + 1$

3 coins: 7¢ $5 + 1 + 1$

2 coins: 26¢ $25 + 1$

2 coins: 35¢ $25 + 10$

5 coins: 14¢ $10 + 1 + 1 + 1 + 1$

Page 66

1. My Grandma baked 2 pumpkin pies. I ate a half of the total amount of pies. How many pie(s) are left?

Circle your answer: 0 (1) 2 3 4 5

A half is one of two equal parts of one whole. If two pies are one whole, I could eat 1 pie which is a half. Another half is left. So, I circle 1.

I found 4 shells. My sister broke a half of the shells. Color these shells red. How many shells are left?

Circle your answer: 0 1 (2) 3 4 5

I got 8 cupcakes. I ate a half of them. Color the cupcakes I ate. How many cupcakes are left?

Circle your answer: 0 1 2 (3) 4 5

Page 61

1. I need money to buy objects. Write the missing numbers.

2 coins: 15¢ $10 + 5$

3 coins: 3¢ $1 + 1 + 1$

4 coins: 40¢ $10 + 10 + 10 + 10$

2 coins: 10¢ $5 + 5$

I bought 5 lollipops for 10 cents. How much did one lollipop cost? 2 cents. $2+2+2+2+2=10$

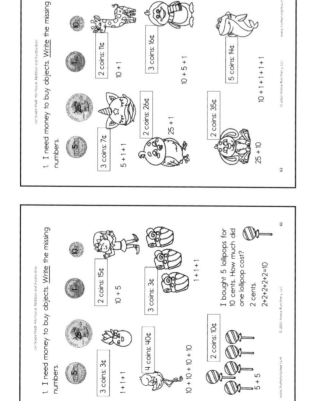
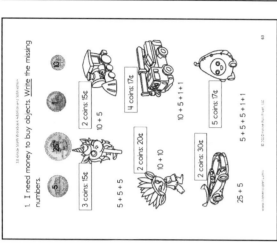

Page 65

1. Write the missing numbers to make the scales balance.

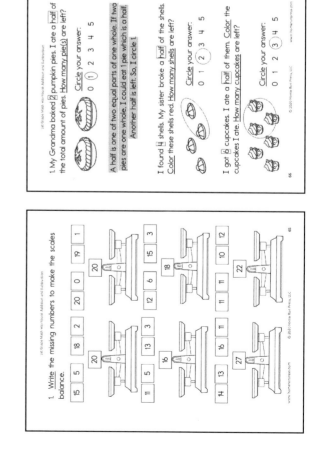
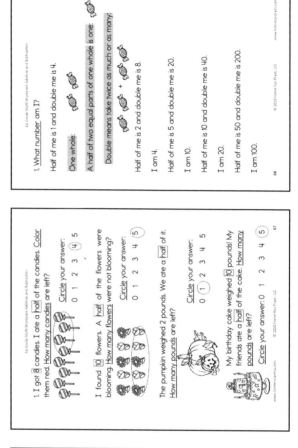

69

1. When you share equally between two elves, both sets of sweets and fruits have the same amount. Count how many for each elf?

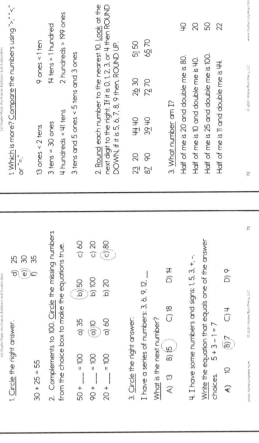

Each elf must have the same amount.

🍌	🍪	🥧
2	5	3

© 2020 Home Run Press, LLC

70

1. I am building a solid slab of rocks: the two rocks next to each other are added to get the number up above. Fill in the missing numbers.

11	
5	6
2	3	3
1	1	2	1

2. Write the numbers in order, from the smallest to the largest.
7, 9, 3, 1, 2, 11, 5, 12, 8, 4, 0, 10, 6
0, 1, 2, 3, 4, 5, 6, 7, 8, 9, 10, 11, 12

3. Complements to 10. Circle the missing numbers from the choice box to make the equations true.

11 + 9 = 20   a) 8   b) 7   c) 9
5 + 15 = 20   a) 10   b) 5   c) 15
7 + 13 = 20   a) 13   b) 1   c) 14
14 + 6 = 20   a) 4   b) 16   c) 6

© 2020 Home Run Press, LLC

71

1. Circle the right answer.
30 + 25 = 55   d) 25   e) 30   f) 35

2. Complements to 100. Circle the missing numbers from the choice box to make the equations true.

50 + ___ = 100   a) 35   b) 50   c) 60
90 + ___ = 100   a) 10   b) 100   c) 20
20 + ___ = 100   a) 60   b) 20   c) 80

3. Circle the right answer.
I have a series of numbers: 3, 6, 9, 12, ___
What is the next number?
A) 13   B) 15   C) 18   D) 14

4. I have some numbers and signs: 1, 5, 3, +, -.
Write the equation that equals one of the answer choices.   5 + 3 - 1 = 7
A) 10   B) 7   C) 4   D) 9

www.homerunpress.com

72

1. Which is more? Compare the numbers using ">", "<", or "=".

13 ones < 2 tens   9 ones < 1 ten
3 tens = 30 ones   14 tens > 1 hundred
4 hundreds < 41 tens   2 hundreds > 199 ones
3 tens and 5 ones < 5 tens and 3 ones

2. Round each number to the nearest 10. Look at the next digit to the right. If it is 0, 1, 2, 3, or 4 then ROUND DOWN. If it is 5, 6, 7, 8, 9 then, ROUND UP.

23 20   44 40   26 30   51 50
87 90   39 40   72 70   65 70

3. What number am I?

Half of me is 20 and double me is 80.   40
Half of me is 10 and double me is 40.   20
Half of me is 25 and double me is 100.   50
Half of me is 11 and double me is 44.   22

© 2021 Home Run Press, LLC

73

1. Solve the problem:
10+7. the sum of the ones and hundreds is 7 + 0 = 7.
A) 5   B) 8   C) 7

2. I start at 0 and count on in twos. Will I say 11? 0, 2, 4, 6, 8, 10, 12
Why? 11 is an odd number.

I start at 0 and count on in twos. Will I say 16? 0, 2, 4, 6, 8, 10, 12, 14, 16
Why? 16 is an even number.

3. Find the value.
853
The sum of the ones and tens is 3 + 5 = 8.
8 - 5 = 3.
The difference between the hundreds and tens is
8 - 5 = 3.
The difference between the hundreds and ones is
8 - 3 = 5.

www.homerunpress.com

74

1. Solve the problems:
I had 8 candies. I gave 4 of them to my sister. How many candies has I left?   8 - 4 = 4 (candies)

There are 10 kids at a playground. I counted 7 boys.   3 girls.
How many girls are there?
10 - 7 = 3 (girls)

My brother bought 11 chocolate cupcakes and 5 vanilla cupcakes. How many cupcakes did he buy in all?   16 cupcakes.
11 + 5 = 16 (cupcakes)

I found 6 easter eggs. My sister found 4 more Easter eggs than I did. My brother found 7 less Easter eggs than my sister. How many Easter eggs did my brother find?   3 Easter eggs.
6 + 4 = 10 (sister)   10 - 7 = 3 (brother)

© 2020 Home Run Press, LLC

75

1. Solve the problems:
I had 15 cupcakes. I ate some cupcakes, and I had 12 cupcakes left. How many cupcakes did I eat?   3 cupcakes: 15 - 12 = 3

My brother has 18 trucks and race cars. 3 of them are trucks. How many race cars does he have?   15 race cars: 18 - 3 = 15

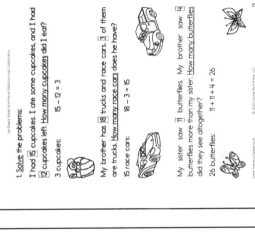

My sister saw 11 butterflies. My brother saw 4 butterflies more than my sister. How many butterflies did they see altogether?   26 butterflies: 11 + 11 + 4 = 26

www.homerunpress.com

76

1. Solve the problems:
I had some cupcakes. I ate 3 cupcakes and I gave 4 cupcakes to my friend. I have 11 cupcakes left. How many cupcakes did I have at first?   18 cupcakes: 3 + 4 + 11 = 18

I had 5 yellow balloons and 5 more red balloons than yellow balloons. My friend had 8 more balloons than I had. How many balloons did my friend have?   18 balloons: 5 + 5 + 8 = 18

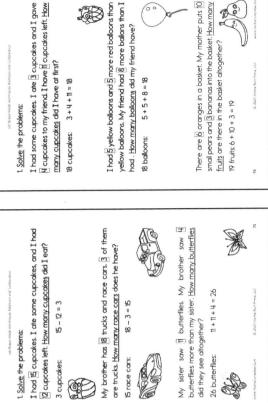

There are 6 oranges in a basket. My mother puts 10 small pears and 3 bananas into the basket. How many fruits are there in the basket altogether?   19 fruits: 6 + 10 + 3 = 19

© 2021 Home Run Press, LLC

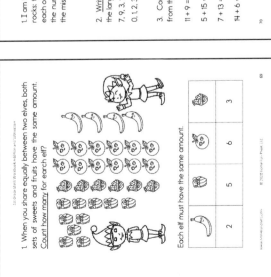
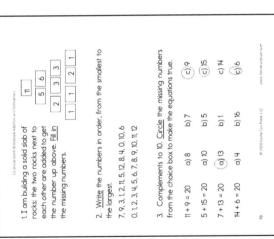

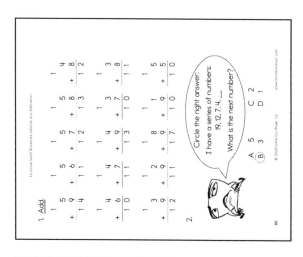

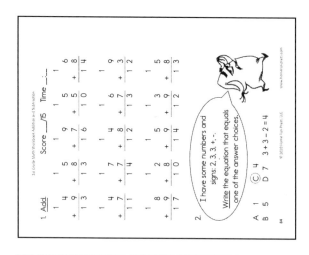

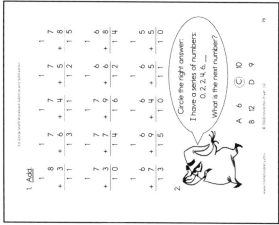

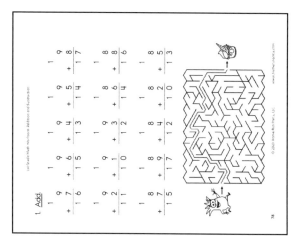

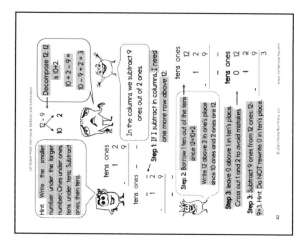

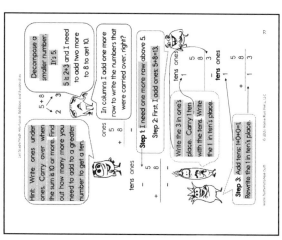

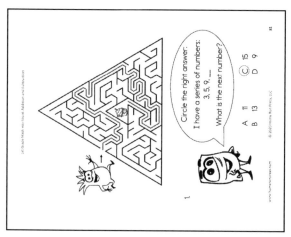

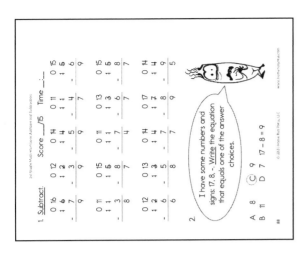

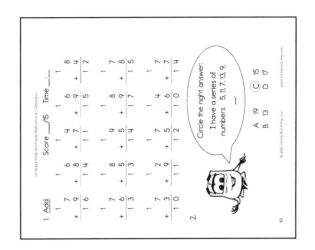

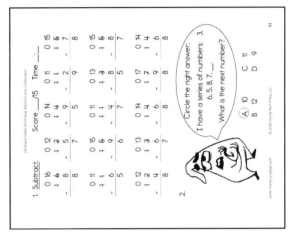

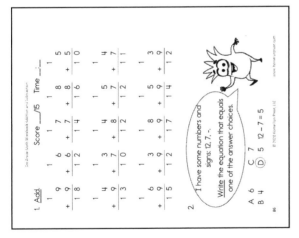

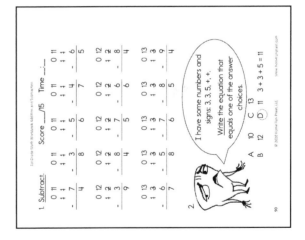

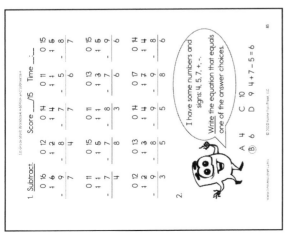

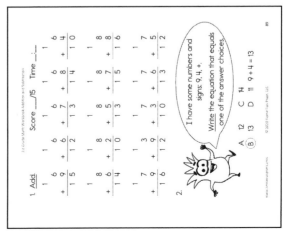

Add

1st Grade Math with Facial Addition and Subtraction

1. Add   Score ___/15   Time ___:___

$$\begin{array}{cccccc} 1 & 1 & 1 & 1 & 1 & 1 \\ 9 & 4 & 7 & 6 & 1 & 8 \\ +\,9 & +\,8 & +\,7 & +\,5 & +\,1 & +\,8 \\ \hline 18 & 12 & 14 & 11 & 5 & 16 \end{array}$$

$$\begin{array}{cccccc} 1 & 1 & 1 & 1 & 1 & 1 \\ 9 & 5 & 9 & 8 & 9 & 8 \\ +\,6 & +\,5 & +\,7 & +\,7 & +\,7 & +\,8 \\ \hline 15 & 10 & 16 & 15 & 16 & 17 \end{array}$$

$$\begin{array}{cccccc} 1 & 1 & 1 & 1 & 7 & 1 \\ 7 & 2 & 9 & 8 & 6 & 8 \\ +\,5 & +\,8 & +\,5 & +\,6 & +\,8 & +\,8 \\ \hline 12 & 10 & 14 & 14 & 13 & 15 \end{array}$$

2. I have some numbers and signs: 6, 7, 11, +, -.

Write the equation that equals one of the answer choices.

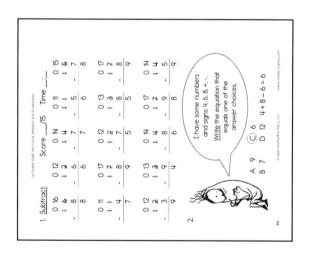

A   13   C   4

B   2   D   16   6 + 7 - 11 = 2

www.hofrunpress.com   © 2020 Home Run Press, LLC   93

Subtract

1st Grade Math with Facial Addition and Subtraction

1. Subtract.   Score ___/15   Time ___:___

$$\begin{array}{cccccc} 0 & 0 & 0 & 0 & 0 & 0 \\ 16 & 12 & 14 & 14 & 11 & 15 \\ -\,8 & +\,2 & +\,4 & +\,1 & -\,5 & -\,7 \\ \hline 8 & 6 & 7 & 7 & 6 & 8 \end{array}$$

$$\begin{array}{cccccc} 0 & 0 & 0 & 0 & 0 & 0 \\ 11 & 17 & 12 & 12 & 13 & 17 \\ +\,4 & +\,7 & +\,2 & +\,3 & +\,8 & +\,8 \\ \hline 7 & 9 & 5 & 5 & 5 & 9 \end{array}$$

$$\begin{array}{cccccc} 0 & 0 & 0 & 0 & 0 & 0 \\ 12 & 13 & 14 & 17 & 14 & 14 \\ +\,2 & +\,3 & +\,4 & +\,2 & +\,4 & +\,1 \\ \hline 9 & 4 & 5 & 8 & 6 & 9 \end{array}$$

2. I have some numbers and signs: 4, 6, 8, +, -.

Write the equation that equals one of the answer choices.

A   9   C   6

B   7   D   12   4 + 8 - 6 = 6

© 2020 Home Run Press, LLC   www.homerunpress.com   94